EINSTEIN'S WIFE

EINSTEIN'S WIFE

THE REAL STORY OF MILEVA EINSTEIN-MARIĆ

ALLEN ESTERSON AND DAVID C. CASSIDY
WITH A CONTRIBUTION BY RUTH LEWIN SIME

The MIT Press
Cambridge, Massachusetts
London, England

This book was set in Adobe Garamond Pro by Jen Jackowitz. Printed and bound in the United States of America.

Library of Congress Cataloging-in-Publication Data

Names: Esterson, Allen, 1936- author. | Cassidy, David C., 1945- author. |
    Sime, Ruth Lewin, 1939- author.
Title: Einstein's wife : the story of Mileva Einstein-Marić / Allen Esterson
    and David C. Cassidy ; essay by Ruth Lewin Sime.
Other titles: Story of Mileva Einstein-Marić | Mileva Einstein-Marić
Description: Cambridge, MA : The MIT Press, [2019] | Includes bibliographical
    references and index.
Identifiers: LCCN 2018029638 | ISBN 9780262039611 (hardcover : alk. paper)
Subjects: LCSH: Einstein-Marić, Mileva, 1875-1948. | Einstein, Albert,
    1879-1955--Marriage. | Einstein, Albert, 1879-1955--Family. | Women
    mathematicians--Biography. | Women mathematicians--Serbia--Biography. |
    Women mathematicians--Germany--Biography. | Mathematicians--Biography. |
    Mathematicians--Serbia--Biography. | Mathematicians--Germany--Biography.
Classification: LCC QC16.E52 E88 2019 | DDC 510.92 [B] --dc23 LC record
available at https://lccn.loc.gov/2018029638
ISBN: 978-0-262-03961-1

10   9   8   7   6   5   4   3   2   1

CONTENTS

ACKNOWLEDGMENTS

The authors express their great appreciation to Alberto A. Martínez and an anonymous referee for their careful reading of the manuscript and for their excellent suggestions.

We wish sincerely to thank Ruth Lewin Sime for her contribution and Gerald Holton for his advice.

We also convey our appreciation to Mihaela Barbarić of the State Archive in Zagreb, Croatia, for providing copies of the documents appearing in figures 1.1 and 1.2, and to Božidar Kereković of the State Archive for permission to publish them.

For their generous assistance in helping us locate and obtain photographs and for their permission to publish them we are very grateful to Or Orith Burla Barnea of The Albert Einstein Archives at The Hebrew University of Jerusalem; to Michael Simonson of the Leo Baeck Institute, New York; to the Niels Bohr Library and Archives, American Institute of Physics, College Park, Maryland; and to the ETH Library Zurich, Photo Archive.

The authors wish to thank Katie Helke and the staff of MIT Press for their expert work and care in the publication of this book.

In addition to the above, Allen Esterson would like to thank David C. Cassidy for his work in getting my chapters into publishable form as the first stage in the production of the final draft.

ABBREVIATIONS

After her marriage Mileva Marić took the name Mileva Einstein-Marić. Thus, before, during, and after her marriage her last name ended in Marić. Solely for simplicity, we use Marić when referring to her by last name.

We use the following abbreviations throughout:

AAAS	American Association for the Advancement of Science
CPAE	Einstein 1987ff. *The Collected Papers of Albert Einstein*, followed by volume number and document or page number. All of the published volumes and their English translations are online at http://einsteinpapers.press.princeton.edu/.
E-M	Einstein and Marić 1992. *The Love Letters*. Many of the letters from and to Mileva Marić were undated or only partially dated. Editors of the volume assigned tentative dates to this correspondence; we have incorporated these dates into the text to ensure chronological clarity, or on some occasions included them in the parenthetical references.
ETH	Eidgenössische Technische Hochschule (Swiss Federal Institute of Technology), until 1911 the Swis Polytechnic Institute, Zurich.

M-KS	Popović 2003. Mileva Einstein-Marić's letters to Helene Kaufler Savić. The volume contains no letters from Kaufler Savić to Marić. The Serbian edition, with original languages, is referenced as Popović 1998.
[ref in original]	When relevant we include in brackets the references cited by other authors we quote, using a format that corresponds to references in our bibliography.
T-G	Trbuhović-Gjurić 1969ff, biography of Mileva Einstein-Marić, followed by year(s) of the edition(s) cited. The 1982 and 1983 editions are identical, as are the 1988 and 1993 editions. References to both the 1983 and 1988 editions are given because new material added to the latter edition caused the pagination to change.
T-P	Troemel-Ploetz 1990, followed by page numbers.

INTRODUCTION

Who was the real Mileva Einstein-Marić, the wife of the famed physicist Albert Einstein? Trained alongside Einstein in physics and mathematics, what role did she play, if any, in the famous papers of her husband, which transformed contemporary physics? Was she an unsung contributor or even a co-author, a sounding board, the top fiddle, a glorified assistant, an unglorified housewife, the one who made it all possible?

What does it even matter? It matters, of course, because Einstein's work and his theories of relativity, quantum theory, and atomic theory formed the foundations of contemporary physics. Without the true story of what really happened—how these fundamental theories came about—we cannot fully understand their historical significance. Moreover, we cannot bestow credit where credit is due for these achievements, especially when we recognize how often the contributions of women scientists, especially the scientific spouses and partners of celebrated male scientists, are overlooked, forgotten, and even suppressed.

Much has been written over the past decades in support of one or more of the above perspectives on Mileva Einstein-Marić, often with fervent certitude This book is an excursion into these heavy waters, a journey guided by the historical urge to "get it right"—or at least to get as close as possible to "right"—for the sake of Mileva and Albert, and for the sake of the history of contemporary physics. Such a journey becomes all the more important in an era of deliberate falsehoods, fake news, and "alternate facts." The ballast of mounting documentary evidence that has become available during the past decades about these two extraordinary people and their work and activities keeps this ship on course.

It is well known that Albert Einstein, the world-famous scientist declared the "person of the century" by *Time* magazine, was married. But until the 1990s, few people knew that he had two wives, the first of whom was trained in mathematics and physics. Einstein's nonscientific second wife, Elsa, did become widely known, especially in the United States, when she accompanied her husband on several well-publicized visits during the 1920s, then settled with him in Princeton in 1933 (she died three years later). The two had married in 1919, the same year that the first public confirmation of Einstein's theory of general relativity, one of the greatest achievements of the twentieth century, brought him worldwide fame.

But it was his first wife, Mileva Einstein-Marić (approximately pronounced *Mar-itch*), who accompanied and supported him intellectually and emotionally throughout the difficult early years of his rise from a beginning physics student in 1896 to the top of his profession by 1914. Little was known about her or his children with her—and there was little interest in finding out—until the discovery in 1986 of her correspondence with Einstein brought Mileva Einstein-Marić to widespread public attention. The letters had been in the possession of the Einsteins' first son, Hans Albert Einstein, and his family in Berkeley, California.

The story of Mileva Einstein-Marić encompasses many facets of great human and scientific interest related to the struggle of women in science as revealed by observing one woman's struggle to realize her dreams for a scientific career and examining a married couple's scientific and personal partnership that unfortunately did not succeed.

Widespread public recognition of the captivating story of Einstein's wife Mileva Einstein-Marić emerged during the years following the publication of the first volume of *The Collected Papers of Albert Einstein* in 1987 (*CPAE* 1). That volume, titled *The Early Years, 1879–1902*, documented young Einstein's youth, education, and early career. Of special interest in the volume was the first publication of fifty-one of the then newly discovered Einstein-Marić letters in the possession of Hans Albert's family. The fifty-one letters covered the period from October 1897 to February 1902, the years during and just after their studies together at the Swiss Federal Polytechnic in Zurich, Switzerland (today the Federal Institute of Technology, or ETH). Subsequent letters have appeared in later volumes. Although she was Serbian, Marić was fluent in German, one of the official languages of the Austro-Hungarian Empire. All of the letters appeared in the original German with an accompanying English translation (*CPAE* 1 [Engl.] and *The Love Letters* [E-M], a more polished translation of the first fifty-one letters, plus three more, through 1903).

That first volume of Einstein's *Collected Papers* also brought to public attention a previously little-known but subsequently highly influential biography of Mileva Einstein-Marić by the Serbian science professor Desanka Trbuhović-Gjurić, *Im Schatten Albert Einsteins: Das tragische Leben der Mileva Einstein-Marić* (In the Shadow of Albert Einstein: The Tragic Life of Mileva Einstein-Marić). Originally published in Serbian in 1969, it went through two German editions and four printings as well as one French translation from 1982 to 1995 (T-G 1969–1995), but it has never been published in English translation. Nonetheless, her biography (through its intermediaries) has helped more than any other work so far to shape the dominant public story of Mileva Einstein-Marić and her personal and scientific relationships with Albert Einstein.

While tracing Mileva Einstein-Marić's life story, Trbuhović-Gjurić (approximately *Tra-bu-ho-vitch Jur-itch*) argued, often (as we will see) without source citation or solid evidence, that Marić was a brilliant mathematician who surpassed Einstein in mathematics, if not physics. Moreover, owing to her expert collaboration with Einstein, Marić was allegedly an unrecognized co-author of her husband's famous relativity paper of 1905. If true, such claims would mean that once again, as so often in the past (and present), the contributions made by the wife of a great man have been sadly overlooked by the public, forgotten by history, and apparently even suppressed by her husband.

On February 18, 1990, nearly three years after the publication of the first volume of Einstein's *Collected Papers*, a session on "The Young Einstein" during the annual meeting of the American Association for the Advancement of Science (AAAS) in New Orleans brought Mileva Einstein-Marić and her marriage with Einstein to widespread public attention (Herschman 1989, 1326). Most of the speakers at the AAAS session presented scholarly reevaluations of Einstein's early biography, cultural environment, and philosophy.[1] But two of the speakers, Senta Troemel-Ploetz and Evan Harris Walker, both of whom were without any prior work on Einstein studies, seized upon the opportunity to create a full-scale interpretation of Marić and her scientific relationship with Einstein for English-speaking audiences. Walker, who died in 2006, was a physicist at the United States Army Ballistic Research Laboratory in Maryland, president of the Walker Cancer Research Institute, which he founded in 1981, and the author of works on parapsychology (e.g., Walker 1979). By 1990 his penchant for pseudoscience was well known (Gardner 1981; Kaiser 2011, 84–86). Troemel-Ploetz is a German linguist and psychotherapist.

Drawing upon Trbuhović-Gjurić and the Einstein-Marić letters for their individual papers, Troemel-Ploetz and Walker forcefully argued the startling conclusion that Mileva Einstein-Marić made substantive contributions to Einstein's early work. In a lengthy paper based on her talk in the journal *Women's Studies International Forum*, for instance, Troemel-Ploetz described Marić as "the woman who did Einstein's mathematics" and as the unjustly overlooked co-author of Einstein's famous 1905 papers. His three main papers, she asserted, "were written in Bern while Albert Einstein was at the Swiss Patent Office and were written together with his wife" (T-P 1990, 419).

In his AAAS paper Walker quoted sentences from Einstein's letters to his future wife from which he concluded, "Einstein's words alone show us Mileva was co-author of the special theory of relativity." And, he surmised, "The most capricious ideas that were the turning point of relativity theory came from Mileva, while much of the overall formalism of the theory was set up by Albert" (Walker 1990).

The sweeping and surprising assertions about Mileva Einstein-Marić and her husband drew enormous public and media interest, as well as an outpouring of books and articles on the apparent injustices suffered by "Einstein's first wife." She seemed to be the most prominent and most blatant example of how history has forgotten, even deliberately so, the contributions female scientific spouses and partners made to the great achievements of male scientists. The early history of contemporary physics thus required immediate correction. Many of the resulting popular works unquestioningly repeated, and even embellished, the claims of their predecessors, in some cases without apparent regard for common standards of nonfiction writing. Over the years what might be called the "Mileva Story" emerged and entered the public domain as the generally accepted account of Mileva Einstein-Marić's unrecognized collaboration with her former husband and her contributions to his work. But at the same time, on closer examination, Einstein scholars have generally rejected most of this story on the basis of the available documentary evidence (e.g., Pais 1994, 1–6, 14–16, 21; Holton 2000; Martinez 2011). As the physics historian Alberto Martínez (2011, 193) wrote, "I want her to be the secret collaborator. But we should set aside our speculative preferences and instead look at the evidence."

Historians' objections to the Mileva Story have gained increasing support in the years since 1990 with the abundance of new documentary material and information made available about Marić, Einstein, and their relationship. Some of this new evidence is now readily accessible in print and online in the volumes of Einstein's *Collected Papers,* published in the original languages and in English translation since 1987. The volumes produced so far extend into 1927 and include more of the Einstein-Marić correspondence; Einstein's correspondence with friends and colleagues regarding his work, papers, and family; letters and records relating to their marriage in 1903, separation in 1914, and divorce in 1919; documents regarding Marić and the Nobel Prize money; and many other relevant materials.

Additional crucial documents beyond the *Collected Papers* include some of Mileva Marić's school grade reports (figures 1.1 and 1.2); her Polytechnic grade reports (Marić, Student Record, see our bibliography), as well as her letters to Helene Kaufler Savić, her close friend and confidante from the early Zurich days, extending from 1899 to 1932. These were first published in their entirety in 1998 in the original German with a parallel

Serbian translation (Popović 1998), and in English translation (Popović 2003, hereafter cited as M-KS).

In addition to the new widely available primary sources, significant new biographies of Einstein have appeared that offer more closely evidence-based accounts of Mileva Marić and her relationship with Einstein (Fölsing 1997; Overbye 2001; Isaacson 2007; Neffe 2007). They are beginning to exert a noticeable impact on the public's understanding and appreciation of her real story.

In view of the mounting primary and secondary material since 1990, it seems appropriate at this time, nearly three decades since the public emergence of the Mileva Story, to reevaluate in this book the many elements of such an important historical assertion on the basis of all of the available evidence, new and old. We proceed in three steps:

1. The first is to provide a biographical sketch of Mileva Marić and her relationship with Albert Einstein, based on our knowledge of current documents, that reveals what she *did* do (David C. Cassidy).
2. The second is to offer an essay placing Mileva Marić within the broader context of the struggle of women to enter science at the turn of the twentieth century, providing also a concise entrée into the literature on this subject (Ruth Lewin Sime).
3. The third is to examine in-depth the origins and assertions of the Mileva Story based on the available evidence, revealing what she did *not* do, as the evidence indicates (Allen Esterson).

This three-step approach also reveals how far one can go wrong when standards for historical and nonfiction writing are not carefully followed. But more importantly, it reveals the very human, real story of a fallible, yet brave and determined young woman who, for various reasons, was not able to fulfill her dreams for the career and marriage she had hoped for. This more realistic and compelling story of Mileva Marić offers a far greater service to her—and to readers' appreciation of her international pioneering role in helping to open science and science education to female students—than any exaggerated or unfounded assertions regarding her activities, as promoted over the past decades.

This book began in 2006 when Allen Esterson, a British independent scholar, undertook the demanding task of carefully analyzing and critiquing, on the basis of the available evidence, the many works in print and in other media regarding Mileva Einstein-Marić and her alleged role in Einstein's achievements. He published most of his findings as essays on his website, http://www.esterson.org. In 2010 he began collecting his essays into a book manuscript and had nearly completed the task when he fell ill in late 2013 and could not continue. The following year Cassidy agreed to help him complete the task, By 2017 Esterson had recovered sufficiently to resume his authorship of what is now part III of this book. It is now entirely his own work. At the same time, Cassidy wrote his biographical account of Mileva and Albert (part I). In addition, Ruth Lewin Sime, the well-known biographer of Lise Meitner, has contributed an eloquent essay on the struggles of women such as Mileva Marić to enter science, which is now part II.

EVIDENCE

The Mileva Story asserts that Mileva Einstein-Marić contributed substantially to Albert Einstein's scientific achievements, especially those of his "miracle year" 1905, and that she should have been rightfully listed as a co-author of one or more of those papers. A claim of this significance regarding scientific achievements of this magnitude demands that we have the most accurate understanding possible of all aspects of these developments, of how and why they came about, and of their role in forming the foundations of today's physics. As has frequently been stated, extraordinary claims require extraordinary evidence.

History and journalism are not exact sciences. The past no longer exists, and we cannot carefully observe what happened or record the actors' thoughts and motivations as they are acting. But we do have a wide range of documents, letters, memories, and contemporaneous remains associated with these developments that can help us learn a great deal about what really happened and why. But they are rarely exhaustive, and important gaps often remain (as they do in this case). In addition, some judgmental evaluation and interpretation of past evidence are always necessary, but we also have well-developed methods and standards that have evolved and gained wide acceptance over centuries for evaluating the clues among the remains and for reporting the results. Most of these are based on common sense. We hasten to add, however, that those wishing to look into the past and to write about a subject such as this, on a nontechnical level, do not need to be trained scholars or journalists, but they should be willing to take the time and effort to familiarize themselves with the widely accepted procedures and standards for handling evidence and writing about it.

The remains of the past that constitute historical "evidence" fall into a wide range of categories displaying a diversity of reliability, usefulness, and accessibility—from contemporary documents, letters, diaries, photographs, and draft papers to recollections and hearsay reports after the fact, even decades later. Like an ancient artifact dug from the sediments of an archeological site, every item of evidence requires careful, critical evaluation regarding its source, its contexts (scientific, cultural, personal), and even the meanings of the words and phrases used at that time.

Those who established and argued the Mileva Story frequently cited hearsay evidence, almost invariably obtained many years after the events, in documenting their claims. Hearsay may be defined as "indirect accounts of what someone allegedly told someone else" (Martínez 2011, 202). In the case of Einstein and Marić the report was often passed down through many "hearers" over many years. According to one list of twenty types of historical sources ranked according to reliability, compiled by Martínez, hearsay evidence repeated over the years ranks at the very bottom of the list, even below memories and interviews conducted after a similar passage of time. Such low reliability is the reason hearsay evidence of any kind is barred from a court of law. No jury or judge should convict a defendant "beyond a reasonable doubt" on the testimony of a witness who says that she heard someone say that the accused said that he committed the crime. This is also the reason why those wishing to build a solid case for an important historical assertion rely instead on more reliable evidence directly from the "crime scene"—verifiable notes, letters, emails, diaries, and drafts created at the time of the event.

To convince oneself of the unreliability of hearsay, one need only refer to the "telephone" or "gossip" game in which people pass a message from one to another and then compare the beginning message with the often widely differing end result. If memory is involved, the result diverges even further from the truth, even for eyewitness accounts. In his book *Memory*, I. M. L. Hunter describes an experiment undertaken by two Cambridge psychologists who, without the knowledge of the participants, recorded a meeting of the Cambridge Psychological Society. Two weeks later they asked all who had attended to write down everything they could recall about the meeting. They found that some

42 percent of the recalled points were substantially incorrect, including happenings that had never taken place at all. "In short, what was recalled was not only fragmentary, but also distorted, and much was recalled which, in fact, had never happened" (Hunter 1957, 160–161). Reporting on more recent experimental work, Micah Edelson et al. (2011, 108) point out: "Human memory is strikingly susceptible to social influences. ... Our memories are often inaccurate. Ubiquitous sources of false recollection are social pressure and interpersonal influence." More generally, as Charles Fernyhough (2012, 130–131) reports in his discussion of research into memory processes: "Findings on rich false memories show that the misinformation is particularly strong when other people, especially family members, are providing the interjected information. Some benefits accrue to collaborative remembering. ... The term *social contagion* is used to describe the process whereby an account of an event incorporates erroneous information provided by other people." To this the science writers Highfield and Carter (1993, 110) add a skepticism of what they describe as "hometown folklore" with its tendency "to exaggerate the claims of local heroes and heroines."

Nevertheless, hearsay and memory, however flawed and unreliable, should not be automatically disregarded. In this case, they convey at least what people associated with the Marić family had come to believe were authentic events. As with some ancient sagas and legends handed down through the ages, what comes to us in the present may harbor a kernel of truth from the past that may be discerned, however vaguely, in the events of the time.

We begin with the life and struggles of Mileva Marić as she faced and overcame the hurdles to women in science and education at the end of the nineteenth century, followed by her studies and her relationship with another physics student, Albert Einstein.

1 MILEVA AND ALBERT

DAVID C. CASSIDY

1 TWO TRAJECTORIES

From the very beginning, Mileva Marić's courage, determination, and ambition were evident as she overcame a physical disability, maintained a steady performance while moving from school to school, became one of the first women to study physics at the high school level in Austria-Hungary, and, especially, left her family and her native Serbia for Switzerland in order to realize her dream.

ENTER MILEVA

Mileva Marić was born on December 19, 1875, in the town of Titel in the predominately Serbian province of Vojvodina on the southern border of Hungary. Hungary was at that time joined with Austria in the Austro-Hungarian Empire, with its capital in German-speaking Vienna. Across the Hungarian border to the south lay the Kingdom of Serbia, which, at Mileva's birth, was still part of the Ottoman Turkish Empire. It gained independence in 1878. To the west lay Croatia, then also part of Hungary.

Most of the Serbs of Vojvodina (*Voy-va-di-na*) were Serbian Orthodox Christians descended from those who had fled north and west into the Hungarian province during the invasions of the Muslim Ottoman Turks in previous centuries. After the founding of the Austro-Hungarian Empire in 1867, Serbian independence and the declining Ottoman Empire accelerated a

Pan-Slavic cultural and political revival. The drive for national unity culminated at the end of World War I in the creation of the independent Kingdom of Serbs, Croats, and Slovenes, which became the Kingdom of Yugoslavia in 1929 (Palmer, Colton, and Kramer 2007). Vojvodina is today an autonomous province of Serbia.

According to biographical accounts (T-G 1983, 1988; Zackheim 2000; Krstić 2004; Milentijević 2015), which often neglect source citations, Mileva's father Miloš Marić was born in 1846 into a peasant family in the town of Kać (*Katch*), Vojvodina. Kać lay near the town of Novi Sad, the political and cultural capital of Vojvodina. In 1694 Serbian merchants crossed the Danube River to found the new city of Novi Sad under pressure from Ottoman Turks and "Germans" immigrating into Hungary from Austria.

Miloš Marić arrived in Titel in 1862 on military service at the age of sixteen. As a buffer against the Ottomans, Hungary had established a Military Frontier along its southern border that was well manned through local military conscription. Sergeant Marić received most of his schooling in a German-speaking battalion military school for non-commissioned officers in Titel. His mastery of German and mathematics opened the door to a career as a civil servant after he left the military (Krstić 2004, 18–20).

Five years after arriving in Titel, Miloš married Marija Ruzić, the daughter of one of the wealthiest landowners in the city. Following the unfortunate deaths of their first two children at early ages, Mileva Marić became their first surviving child. Two more children followed: a sister, Zorka, and a brother, Miloš Jr. When baby Mileva began to walk, it was discovered that she had a congenital dislocation of her left hip, a condition passed down through her mother's side of the family. Mileva's sister Zorka reportedly also suffered from this condition (Milentijević 2015, 21). It left Mileva with a noticeable limp and one of the lifelong burdens she would have to bear.

PRIMARY SCHOOL

As the Ottoman threat subsided, Hungary dissolved its Military Frontier, and Miloš returned to civilian life in the year of Mileva's birth. Apparently well endowed financially through his dowry and a civil service job at the court of justice in nearby Vukovar, he purchased a large estate near his hometown of Kać that included farmland tilled by local peasants (Milentijević 2015, 20). Miloš later built a large two-story summer home on the Kać estate, which became thereafter a favorite summer retreat for family and friends. In 1877 he was appointed clerk at the district court of justice in the town of Ruma, Vojvodina (Krstić 2004, 21). Rather than settling at Kać, the family moved to Ruma, about 30 kilometers south of Novi Sad, where Miloš began his job as clerk and Mileva, aged six, began school in 1882.

Upon completing four years of primary education in Ruma, Mileva enrolled in fall 1886 for the first year of middle school in the Serbian Higher Girls School in Novi Sad, "higher" referring to more advanced education. Her very supportive father arranged for room and board with a widow in the capital city (Milentijević 2015, 24).

While in primary school, Mileva encountered the unfortunate teasing of her classmates because of her limp. "Her understandable reaction to this ridicule was to withdraw into herself and avoid other children," writes Milentijević (2015, 23; see also Krstić 2004, 22). She was often taciturn in later years. Mileva reportedly showed early academic promise in Ruma and Novi Sad. According to Trbuhović-Gjurić, a Ruma teacher told her father: "Take good care of this child! She is a rare [*seltsames*] phenomenon" (T-G 1983 and 1988, 21). Dord Krstić wrote: "At the end of [the Novi Sad] school year she received the grade of 'excellent' in all subjects"; in addition, in 1961 a former classmate told Krstić that Mileva was the brightest girl in the class, "She always knew everything" (Krstić 2004, 22). These authors offer no further details, but Mileva's father seemed determined that she would receive an education as far as her abilities would take her—this despite the prevailing hurdles for women in education

European students normally attended four or five years of primary school. Mileva attended four years at Ruma. Those bound for university and professional careers then attended a nine-year academic middle and high school, often called a gymnasium (pronounced *ghim-nah-ze-um*), following the Latin usage for "school." This generally resulted in fourteen years of schooling before arriving at the university, instead of the twelve in American schools. The gymnasia of the day were typically segregated by gender and were usually attended only by males in countries such as Austria-Hungary that excluded women from higher education. Most offered an academic curriculum emphasizing classical languages and literature. After successfully completing the gymnasium, those wishing to go on to higher education had to pass, in addition to the last year's final examinations, further examinations for the Matura (Abitur in Germany), a certificate granting admission to higher education (Clark 2008, 167–176).

MIDDLE SCHOOLS

Since Austria-Hungary did not permit women to enroll as students in its universities or to receive a degree, there was no gymnasium for Mileva to attend after completing primary school. Instead, like most girls, she attended a girls' middle school, the

Serbian Higher Girls' School in Novi Sad, which paralleled the early years of a male gymnasium. That level of education for women was apparently deemed sufficient for their future role in starting their sons on a route to the higher education and professional careers barred to their mothers.

After just a year at the Serbian Girls' School, Mileva transferred in fall 1887 to another middle school, the Royal Lower Secondary School in Sremska Mitrovica on the Sava River not far from her family's home in Ruma ("lower" referring to the gymnasium grades) (*CPAE* 1, 380–381). Milentijević (2015, 25) attributes this move to her desire to attend "a school better suited to challenging her academic talents." Three years later, having completed four years of "middle school," Mileva was now fourteen, and, writes Ruth Lewin Sime (1996, 6), "public school for girls was over at age fourteen." Rather than spending "the next few years helping at home, sewing, and daydreaming of marriage" (Sime 1996, 7), as did most Austro-Hungarian girls of that era, Mileva was determined to continue her education at an actual gymnasium in order to prepare for higher education. She and her resourceful father found a solution. Twelve years earlier the Kingdom of Serbia, to the south, had gained independence from the Ottoman Empire, and Serbia did allow women to enroll in its universities, which meant it also admitted them to its gymnasia.

GYMNASIUM

Mileva headed just across the border to the town of Šabac, Serbia, farther down the Sava River but still west of Belgrade, the Serbian capital. There, in fall 1890, she enrolled in the fifth of nine classes (or annual "forms") at the Royal Serbian Gymnasium. She remained there for the following school year, 1891–92. But during that year the Marić family experienced a major change. In December 1891 the government issued a decree appointing Miloš Marić to a new position, as of May 1892, as "the lower judicial official at the High Court of Justice in Zagreb," the capital of Croatia, which was then a province of Austria-Hungary (Krstić 2004, 27–28, citing the decree). The Marić family moved from Ruma to Zagreb in May, a move that required Mileva to withdraw from the Šabac gymnasium before she could complete her sixth gymnasium year.

Once settled in Zagreb, Mileva's father, a civil servant at the High Court, managed to obtain permission for Mileva to enroll as a private student, and without the necessity of paying fees, in the public male Royal Upper Gymnasium (also called the Royal Classical Gymnasium) in Zagreb. After passing the entrance exams, Mileva enrolled in the gymnasium in fall 1892. In the following year she even received a stipend (Krstić 2004, 28–29).

The Croatian State Archive in Zagreb kindly provided Allen Esterson with Mileva's school records, including her grade transcripts, for the two years she attended the Zagreb Gymnasium. (The grade transcripts are reproduced in figures 1.1 and 1.2 and in translation in appendix A.) The records indicate that she was admitted in 1892 into the sixth class of the gymnasium, which they designated as "VI a," *a* possibly referring to a second simultaneous class at the same level.

The curriculum at the Zagreb gymnasium was heavily weighted toward languages. In addition to the two classical languages, standard at a "classical" gymnasium, Mileva enrolled in German and Croatian language classes—a remarkable four language classes in all. Of the roughly forty students in her class in the first year and sixty students in the second year (from her school records), two were private students in the public school, except for only one private student (Mileva) in her second

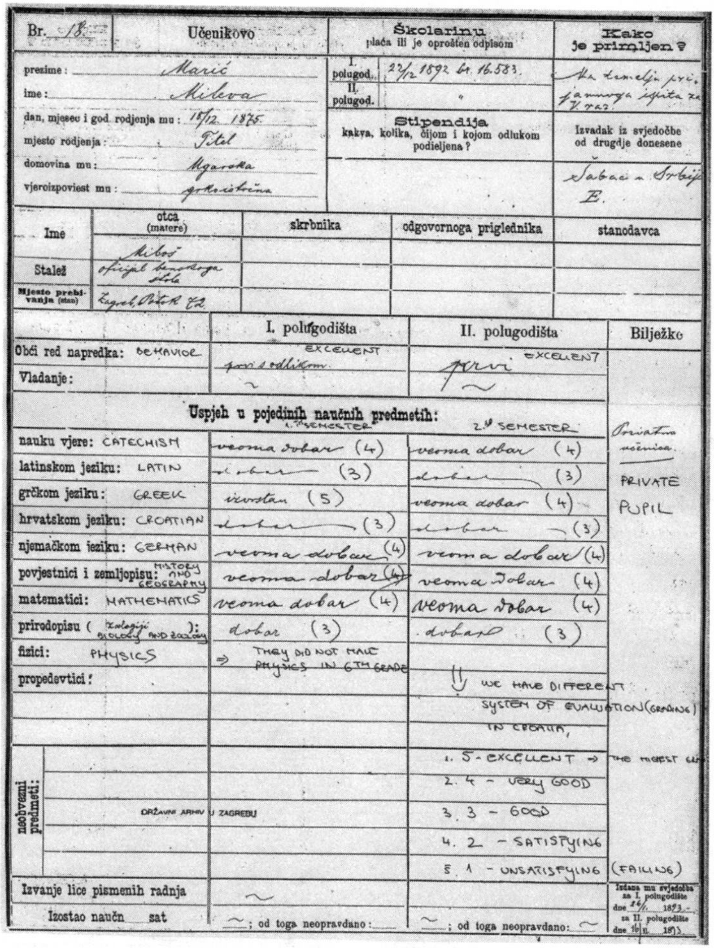

Br. 18 | Učenikovo | Školarinu plaća ili je oprošten odpisom | Kako je primljen?

prezime: Marić
ime: Mileva
dan, mjesec i god. rodjenja mu: 19/12 1875
mjesto rodjenja: Titel
domovina mu: Ugarska
vjeroizpoviest mu: grkoistočna

I. polugod. 22/12 1892 br. 16.583
II. polugod. "

Stipendija kakva, kolika, čijom i kojom odlukom podieljena?

Na temelju privatnog ispita za V. raz.

Izvadak iz svjedočbe od drugdje donesene
Šabac u Srbiji
E.

	otca (matere)	skrbnika	odgovornoga priglednika	stanodavca
Ime	Miloš			
Stalež	oficijal banskoga stola			
Mjesto prebivanja (stan)	Zagreb, Petrik 72			

	I. polugodišta	II. polugodišta	Bilježke
Obći red napredka: BEHAVIOR	prvi s odlikom. EXCELLENT	prvi EXCELLENT	
Vladanje:	~	~	
Uspjeh u pojedinih naučnih predmetih:	1.st SEMESTER	2.nd SEMESTER	Privatna učenica. PRIVATE PUPIL
nauku vjere: CATECHISM	veoma dobar (4)	veoma dobar (4)	
latinskom jeziku: LATIN	dobar (3)	dobar (3)	
grčkom jeziku: GREEK	izvrstan (5)	veoma dobar (4)	
hrvatskom jeziku: CROATIAN	dobar (3)	dobar (3)	
njemačkom jeziku: GERMAN	veoma dobar (4)	veoma dobar (4)	
povjestnici i zemljopisu: HISTORY AND GEOGRAPHY	veoma dobar (4)	veoma dobar (4)	
matematici: MATHEMATICS	veoma dobar (4)	veoma dobar (4)	
prirodopisu (zoologiji): BIOLOGY AND ZOOLOGY	dobar (3)	dobar (3)	
fizici: PHYSICS	⇒ THEY DID NOT HAVE PHYSICS IN 6TH GRADE		
propedevtici:		⇓ WE HAVE DIFFERENT SYSTEM OF EVALUATION (GRADING) IN CROATIA,	
neobvezni predmeti:	DRŽAVNI ARHIV U ZAGREBU	1. 5 - EXCELLENT → THE HIGHEST ...	
		2. 4 - VERY GOOD	
		3. 3 - GOOD	
		4. 2 - SATISFYING	
		5. 1 - UNSATISFYING (FAILING)	
Izvanje lice pismenih radnja	~		Izdana mu svjedočba za I. polugodište dne 14/1. 1893.
Izostao naučn sat	~; od toga neopravdano:	~; od toga neopravdano: ~	za II. polugodište dne 16/II. 1893.

Figure 1.1
Grade report for Mileva Marić, class VI a, Royal Classical Gymnasium, Zagreb, Croatia, 1892–1893. Translations and remarks by archivist Mihaela Barbarić. See appendix A. Courtesy of State Archive in Zagreb, HR-DAZ6–102, I. Klasična Gimnazija u Zagrebu, Glavni Imenik Za VI a. Razred Sa Šk. God 1892–1893, call no. 25353.

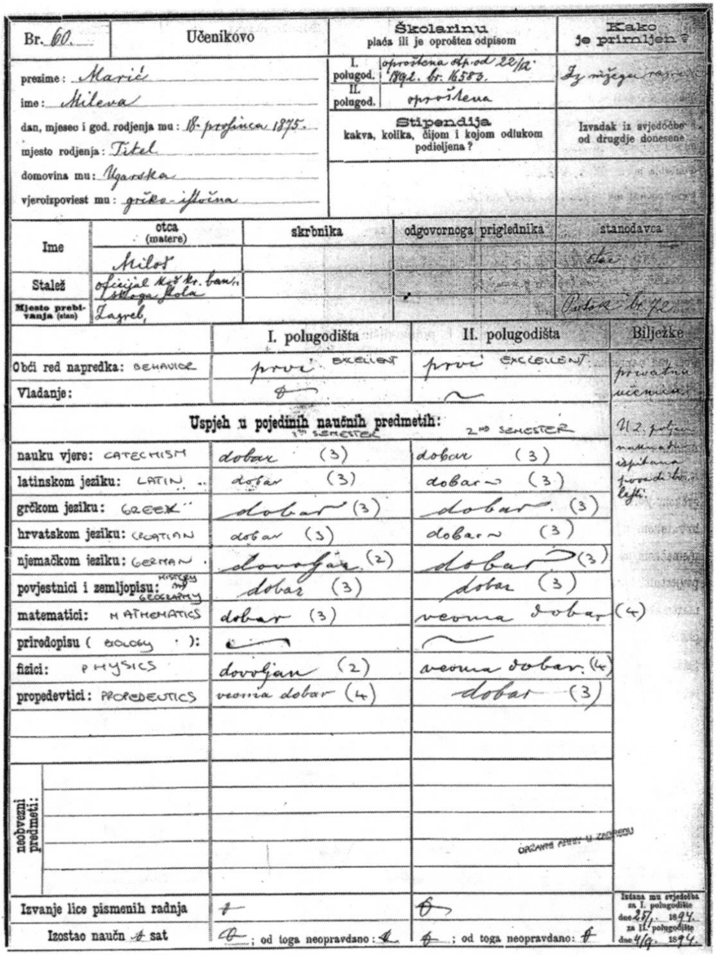

Br. 60.	Učenikovo	Školarinu plaća ili je oprošten odpisom	Kako je primljen?
prezime: Marić ime: Mileva dan, mjesec i god. rodjenja mu: 19. prosinca 1875. mjesto rodjenja: Titel domovina mu: Ugarska vjeroizpoviest mu: grčko-istočna		I. polugod.: oproštena odp. od 22/12. 1892. br. 16583. II. polugod.: oproštena Stipendija kakva, kolika, čijom i kojom odlukom podieljena?	Iz mužega razreda Izvadak iz svjedočbe od drugdje donesene

	otca (matere)	skrbnika	odgovornoga priglednika	stanodavca
Ime	Miloš			otac
Stalež	oficijal kr. kr. ban. stola ...			
Mjesto prebivanja (stan)	Zagreb			Petrin. br. 72

	I. polugodišta	II. polugodišta	Bilježke
Obći red napredka: BEHAVIOR	prvi EXCELLENT	prvi EXCELLENT	privatna učenica
Vladanje:	—	—	
Uspjeh u pojedinih naučnih predmetih:	1ST SEMESTER	2ND SEMESTER	U 2. polg. naknadno ispitana poradi bolesti.
nauku vjere: CATECHISM	dobar (3)	dobar (3)	
latinskom jeziku: LATIN	dobar (3)	dobar (3)	
grčkom jeziku: GREEK	dobar (3)	dobar (3)	
hrvatskom jeziku: CROATIAN	dobar (3)	dobar (3)	
njemačkom jeziku: GERMAN	dovoljan (2)	dobar (3)	
povjestnici i zemljopisu: HISTORY AND GEOGRAPHY	dobar (3)	dobar (3)	
matematici: MATHEMATICS	dobar (3)	veoma dobar (4)	
prirodopisu (BIOLOGY):	—	—	
fizici: PHYSICS	dovoljan (2)	veoma dobar (4)	
propedevtici: PROPEDEUTICS	veoma dobar (4)	dobar (3)	
neobvezni predmeti:			
Izvanje lice pismenih radnja	—	—	Izdana mu svjedočba za I. polugodište dne 27/1. 1894. za II. polugodište dne 4/9. 1894.
Izostao naučn. sat	—; od toga neopravdano: —	—; od toga neopravdano: —	

Figure 1.2
Grade report for Mileva Marić, class VII, Royal Classical Gymnasium, Zagreb, Croatia, 1893–1894. Remarks by the archivist. See appendix A. Courtesy of State Archive in Zagreb, HR-DAZ6–102, I. Klasična Gimnazija u Zagrebu, Glavni Imenik Za VII. Razred Sa Šk. God 1893–1894, call no. 25353.

semester. It is not known if the other private students were girls as well.

Mileva held her own ground at the male gymnasium. Her semester grades for 1892–93, her first year there, were nearly all "very good" (corresponding to a US/American B), including for mathematics. She received for Greek the grade "excellent" (A), her only "excellent" while at the school. (See figures 1.1 and 1.2, and appendix A.) But then something happened. In Mileva's second year at the Zagreb gymnasium, 1893–94, most of her grades dropped a notch, and two landed near the bottom at "satisfactory" (D). We can only surmise what happened. Two possibilities come to mind. One is that she became ill that semester and her grades suffered. Trbuhović-Gjurić writes that when she left the school at the end of her second year she was in ill-health. This was reportedly the reason she postponed her final exams for 1894 from June to September (T-G 1983, 28).

A second possibility derives from the special permission she obtained from the education ministry to enroll in physics that year (T-G 1983, 26). The school permitted only boys to enroll in physics. The audacity of one of the few (or only) girls in the school to utilize a higher authority to force her way into the male world of physics could have awakened resentments and even harassment from the other students. Her performance, or the assessment of it by her male teachers, dropped across the board (except in Latin). Her previous "very good" (B) in mathematics and German dropped to "good" (C) and "satisfactory" (D), respectively; and her first-semester grade in physics was only a barely passing "satisfactory" (D). Whatever the nature of the situation, she persevered and even managed to thrive as best as possible. She was indeed a survivor. The education ministry did not issue its official permission for her to attend the physics class until February 14, 1894, about the time the second semester started (Krstić 2004, 29, quoting the document). With any opposition now silenced, her grades suddenly improved. As shown in the transcripts and in appendix A, Mileva's final semester grades for the delayed exams on September 4, 1894, were all "good" (C), with the exception of mathematics and physics. Her mathematics grade jumped from "good" (C) to "very good" (B), and her physics grade jumped two whole notches, from D to B.

Mileva's best grades were in math and physics that semester, but they were still one notch below "excellent" (A). Some authors have reported these results in confusing, even misleading, terms. Trbuhović-Gjurić, who likely saw Mileva's student records, wrote rather ambiguously, "She had passed the final examination of the seventh class in September 1894 with the best grades [*mit den besten Noten*] in mathematics and physics" (T-G 1983 and 1988, 26–28; see also Frize 2009, 274). More precisely, according to the grade report, these were *her* best grades for that semester.

ZURICH HIGHER GIRLS' SCHOOL

After taking the 1894 exams in Zagreb, Mileva traveled with her father to Zurich in Switzerland, which was among the few European countries that fully admitted women to higher education and degrees. The University of Zurich began admitting women students in 1864; the other Swiss universities and institutes, including the Zurich Polytechnic, followed in 1872. Many wealthy and aristocratic women who could afford the expense traveled to France and Switzerland from other nations that barred women from higher education: those with German language skills (such as Mileva Marić) generally went to German-speaking northern Switzerland, while those with French skills (such as Marie Skłodowska, later Curie) went to France (Clark 2008, 184–189; Kien and Cassidy 1984). The University of Vienna, in contrast, began admitting women only in 1897. Although the gymnasia there continued to exclude women, women were permitted to enter the university if they could somehow pass the Matura exams. After intensive private instruction and study, the Austrian physicist Lise Meitner managed to pass the Matura examinations administered at a male gymnasium in Vienna in 1901. She entered the University of Vienna the same year, and in 1906 she became the second woman awarded a doctorate in physics by that university (Sime 1996, 8–9). The University of Zurich had awarded its first doctorate to a woman, a Russian medical student, in 1867 (Clark 2008, 185).

Mileva enrolled in November 1894 for her last years of preparatory schooling at Zurich's Higher [advanced] Girls' School (Höhere Töchterschule). Her father found room and board for her in the home of a Miss Bächtold; hers was one of the many homes that served as rooming houses for Zurich students. But all was not well in the Marić family. Krstić writes that Mileva's father suffered from rheumatism, and in June 1895, just as Mileva completed her first year in Zurich, he took paid sick leave from his position at the High Court in Zagreb. In January 1896 he retired at the age of forty-nine with a pension nearly equal to his annual salary, but it did not include the housing allowance he had received (Krstić 2004, 31). Soon after taking sick leave, Miloš moved his wife and remaining children back to Vojvodina from Zagreb and purchased a house in Novi Sad, where they settled. From then on, Mileva regarded Novi Sad as home. Her father died in 1922.

Unlike the Zagreb gymnasium, the Zurich girls' school not only prepared and encouraged its female students to take the Matura examinations, but it also expected them to pass, thereby qualifying them for direct admission to higher education.

According to a facsimile of Mileva's entrance certificate to the Zurich girls' school, published by Trbuhović-Gjurić (1983 and 1988, 32), she passed the entrance exams to the school, but the school still required her to take private instruction in French and "perhaps [*eventuell*]" in history, geography, zoology, and botany. She had taken a year of biology and zoology in Zagreb, but French was not offered. Her French classes in Šabac, followed by reported private lessons, were apparently not sufficient (Krstić 2004, 27). As one of Switzerland's four national languages, French was required of all Matura students (the other three languages were/are German, Italian, and Romansh). Although Mileva had studied history and geography, they were probably oriented toward Serbia and Croatia, rather than Switzerland.

ENTERING THE POLYTECHNIC

Mileva passed the Matura examinations in the spring of 1896. She took the exams, not in Zurich, but at the Swiss Federal Medical School in Bern, apparently because she intended at that point to study medicine (Marić, Student Record; T-G 1988, 31–33; Krstić 2004, 21–32). Her grades at the Zurich girls' school and on the Matura examinations have not been found. Trbuhović-Gjurić (1983, 33) provides an extensive list of the school's subjects with instructors, including those in mathematics and physics, probably from a school publication, but no grades.

However, we do have one set of examination results for Mileva in 1896. After attending courses in the University of Zurich's medical program during the summer semester 1896, she decided to study physics and mathematics at the Swiss Polytechnic Institute, or Zurich Polytechnic. She began to do so in the winter semester. (The academic year consisted of winter and summer semesters.) The former Polytechnic (now ETH) has released online her complete student records, which include her

Matrikel, or semester course reports, and her leaving certificate (*Abgangs-Zeugnis*) (Marić, Student Record). Trbuhović-Gjurić published facsimiles of the first page of the Matrikel and the complete, one-page leaving certificate, listing her courses and her average grade for each (T-G 1983, 56–57; 1988, 60–61). The former page shows that Mileva submitted a "federal medical Matura certificate" (*Eidg. mediz. Maturitätszeugnis*) for admission. For some reason, perhaps related to her Matura exams at the medical school, the Polytechnic officials required her to take the entrance examinations in Mathematics and in Descriptive Geometry, but in none of the other eight subjects. Her grades on these exams (also listed in appendix A) ranged from 3.5 to 5 on a scale of 1 to 6 (6 being the highest), resulting in an average of 4.25 out of 6. Mileva passed the exams and immediately switched from medicine at the university to physics and mathematics at the Polytechnic. (In 1911 the Polytechnic became what it is called today, the Swiss Federal Institute of Technology [Eidgenössische Technische Hochschule], or ETH.)

Mileva entered the Swiss Polytechnic in October 1896. It is not known if she intended to pursue a career as a university physics professor or as a gymnasium teacher of physics and mathematics. Whatever her goal, she was surely aware that her possible professions—medicine and university or gymnasium physics—exhibited strong gender discrimination and that few women had succeeded in these professions. Yet she was undeterred, even as she encountered hurdles in her studies, and for this, writes Einstein biographer Walter Isaacson (2007, 137), she deserves our admiration: "Nowadays, when the same issues still reverberate across a century of time, the courage that Marić displayed by entering and competing in the male-dominated world of physics and math is what should earn her an admired spot in the annals of scientific history."

Mileva enrolled in the Zurich Polytechnic's Department VI, the School for Mathematics and Science Teachers, in Section VI A (the mathematics section, which included physics and astronomy). This section offered a four-year program toward the diploma for specialized teachers in these subjects. (Section VI B encompassed the other sciences.) There was no purely academic physics program. That year 841 students enrolled in the Zurich Polytechnic, of whom 42 were in Department VI; of these, 9 were female students, 4 of whom, including Mileva Marić, were in Section VI A. But only one, Mileva, was in the first-year class, and she remained the only woman in her class for the rest of her studies at the Polytechnic (Stachel 2002, 30). In addition, she was one of only two students majoring in physics in the first-year class. The other four students majored in mathematics. They were Marcel Grossmann, Louis Kollros, Jakob Ehrat, and Louis-Gustave du Pasquier (*CPAE* 1, doc. 42). Among that entering class Mileva first encountered the only other first-year physics student that year, Albert Einstein. Albert was at that time seventeen and a half years old; Mileva was now nearly twenty-one.

ENTER ALBERT

Einstein's path to the Zurich Polytechnic was much shorter and more direct that Mileva's. He was born in Ulm in southwestern Germany on March 14, 1879, to non-observing Jewish parents, Hermann and Pauline Einstein. His sister Maria (nicknamed Maya or Maja) was born on November 18, 1881. She later attended the universities of Berlin and Bern, receiving a doctorate in Romance languages at the University of Bern in 1909 (*CPAE* 1, 389).

A year after Albert's birth, the Einstein family moved to Munich, the capital of Bavaria in southeastern Germany, where

Hermann became a partner in the electrical engineering firm of his brother Jakob, a certified engineer. In 1885 the brothers jointly founded a new electrical engineering company in Munich, Einstein & Co., which made electric motors and dynamos. It became a source of Einstein's abiding interest, despite his theoretical research, in practical electro-technology (Pyenson 1985, 35–57).

Albert entered primary school in Munich at the statutory age of six. All public schools in Munich had a religious orientation. Since the last Jewish school had closed earlier, Albert entered the local Roman Catholic primary school in October 1885, receiving private Jewish religious instruction at home (*CPAE* 1, 370). Brief information about his early academic performance at the school comes from a letter that his mother wrote to her sister Fanny, dated August 1, 1886, when Albert was seven, in which she wrote: "Yesterday Albert got his grades, once again he was ranked first, he got a splendid report card" (*CPAE* 1, doc. 2).

EARLY MATH AND SCIENCE

In 1888, at the early age of nine and a half, Albert entered Munich's Luitpold Gymnasium, a classical gymnasium where Greek and Latin held pride of place in the school curriculum (Fölsing 1997, 19–20). It is very unfortunate that the school's records were destroyed during World War II. The only information we have about Albert's grades comes from the rector of the successor school who reported in 1929 that Albert received grades from 1 to 2 (1 being the highest, 4 the lowest) for Latin and 1 to 3 in Greek (*CPAE* 1, lx, n. 46), the subjects apparently of most interest to the rector. Surprisingly, Einstein eventually became acquainted with a total of six languages: German (including Bavarian dialect), Greek, Latin, Italian, French, and finally English.

According to a memoir by Einstein's sister, outside of school Albert exhibited signs of precocious talent in science and, especially, in mathematics (*CPAE* 1, lxi). In 1889 the Einstein family invited a local medical student, Max Talmud (later Talmey), to their home for a weekly midday meal on Thursday, a custom that continued for some five years (Isaacson 2008, 18). Talmey was twenty-one when he first accepted the Einsteins' hospitality. He records that young Albert showed a particular inclination toward physics, so he brought him popular physics books to read. When he observed that Albert showed a great predilection for mathematics, he also brought him a textbook on geometry for self-study. Having worked through the book in a few months, the boy began studying a higher level of mathematics by himself. As Talmey (1932, 164) reported, "Soon the flight of his mathematical genius was so high that I could no longer follow."

Einstein's recollection of this period, published in 1949, is consistent with Talmey's. He described his excitement when his father showed him a "wonder" at the age of four or five, a magnetic compass. "That this needle behaved in such a determined way did not at all fit into the nature of events. ... This experience made a deep and lasting impression upon me" (Einstein 1979, 9). It was an impression that prefigured his interest in electromagnetic fields a decade later. At the age of twelve, Einstein continued, he experienced "a second wonder." It was "a little book dealing with Euclidean plane geometry," perhaps the one provided by Talmey. Einstein further recalled that between ages twelve and sixteen:

> I familiarized myself with the elements of mathematics, including the principles of differential and integral calculus. ... This occupation was, on the whole, truly fascinating; there were peaks whose impression could easily compete with that of elementary geometry—the basic idea of analytical geometry, the infinite series, the

concepts of derivative and integral. I also had the good fortune of getting to know the essential results and methods of the entire field of the natural sciences in an excellent popular exposition, which limited itself almost throughout to qualitative aspects (Bernstein's *Popular Books on Natural Science*, a work of five or six volumes), a work that I read with breathless attention. I had also studied some theoretical physics when, at the age of seventeen, I entered the Polytechnic Institute of Zürich as a student of mathematics and physics. (Einstein 1979, 15)

Although Albert had "familiarized" himself with differential and integral calculus, and his own studies of physics and mathematics reportedly took him beyond the level of his schoolmates (Talmey 1932, 162–164; Fölsing 1997, 21–24), his knowledge of calculus was apparently not at a high level. His first-year grades for this subject at the Polytechnic were 4.5 and 5 out of a possible 6; and his early grades on mechanics and physics were both 5 (see appendix C).

Albert did not thrive in the uncongenial atmosphere of the Munich gymnasium. Following a brief spell of religiosity, his mounting rebellion against religion fostered an antipathy toward all forms of dogma and authority, especially the Prussian influence on his Munich school. According to Maria, "The military tone of the school, the systematic training in reverence for authority that was supposed to accustom pupils to military discipline, was particularly unpleasant. ... Psychologically depressed and nervous, he sought a way out" (*CPAE* 1, lxiii).[1]

ITALY

Albert soon found an escape. In 1894, for economic reasons, the Einstein brothers moved their engineering firm, together with their families, from Munich to Pavia and then to Milan in northern Italy. But Hermann left his son in Munich with relatives in order to complete his schooling. In late December 1894, Albert, now fifteen and in his seventh gymnasium year, quit school and headed to Italy to join his family. But before he quit, he obtained from his mathematics teacher a statement reportedly "affirming that his extraordinary knowledge of mathematics qualified him for admission to an advanced institution" (Frank 1947, 16).

Albert intended to pursue higher education. Lacking the Abitur (Matura) certificate, he planned to study on his own to take the entrance examinations to the Zurich Polytechnic in October 1895. If he passed the exams, he would not need the Matura. During the next nine school-free months in Italy he not only studied for the exams but also learned Italian and wrote his first essay on physics, "On the Investigation of the State of the Ether in a Magnetic Field." It is included in the first volume of his *Collected Papers* (*CPAE* 1, doc. 5). Still, Albert needed special permission from the Polytechnic's director to take the exams, which he obtained with the help of a family friend, since he was two years below the stipulated minimum age of eighteen for admission.

Einstein took the exams as planned. He failed. We do not have the exam grades. Recollections by Einstein and others do not attribute the failure to his performance on the math and science exams. If we can believe later reports, his physics grade was high, but he failed on the general part of the entrance exam, which tested knowledge in political history, literature, languages, and descriptive natural sciences (*CPAE* 1, 10–11; Einstein 1956a, 9; Fölsing 1997, 37).

THE AARAU SCHOOL

Following the personal advice of the Polytechnic's director, young Einstein continued his secondary school education at the

Aargau Cantonal School in nearby Aarau, Switzerland, during the academic year 1895–96. (A Swiss province is called a canton.) There he stayed with the family of Jost Winteler, a teacher of Greek and history at the school. An "Entrance Report" from the school, apparently based on the school's entrance exams, coincides with recollections of Einstein's performance on the Polytechnic exams. On a grade scale that was the reverse of the Polytechnic's, Albert achieved a 2 (1 being the highest, 6 the lowest) in mathematics and physics, but 3s in Italian, Swiss natural history, and history. His performance in two other subjects was noted: "must do catch-up work" in chemistry, and "has large gaps" in French (*CPAE* 1, doc. 8).

Boarding in the teacher's home, Albert thrived in the free atmosphere of democratic Switzerland and in the academic challenges of the preparatory school. We have all of his grades from Aargau, but the school entrance grades and the semester grades as well have caused confusion because of the reversed grading scale during his first two semesters. The school shifted to the Polytechnic's system beginning in Albert's third, and final, semester (see appendix B). As a result, his seemingly poor grade of 1 in mathematics appeared to jump abruptly to 6 the following semester; grades of "1–2" in physics in the first semester jumped to "6–5" in the second; 1 and 2 in chemistry became 5; as did a 2 in French (*CPAE* 1, doc. 10).

By summer 1896 Albert had caught up sufficiently in most subjects (*CPAE* 1, doc. 19). For his final exam grades at the cantonal school, Albert achieved the top grade of 6 in algebra and geometry, "5–6" in physics, and 5s in chemistry, descriptive geometry, history, and Italian, but only 3 in French (*CPAE* 1, doc. 19; see appendix B). He then took the Matura examination which included written and oral portions on most of these same subjects. He passed with a grade average of 5 1/3 on the seven subject examinations. Overall, according to the Einstein editors (without source citation), his grades were the highest of the nine candidates who took the Matura exams at the school, this despite being only seventeen (*CPAE* 1, ed. note, p. 25 and docs. 21–27). At that young age Albert was now qualified to enter the Zurich Polytechnic without needing to retake the entrance examinations. Five other Aargau Cantonal School students joined him and Mileva Marić at the Polytechnic that fall.

At the crossroads of central Europe, German-speaking Zurich in the north of neutral Switzerland was a hub of intellectual, cultural, and political ferment in the late nineteenth century. Its democracy, schools, and two institutions of higher learning open to both genders and all ethnic groups made it a magnet for ambitious students, especially those shut out of higher education in their home countries.

Physics teaching and research at Zurich's Swiss Federal Polytechnic took place within the newly constructed and well-equipped Institute for Physics, which was divided into two sub-institutes, each headed by a single full professor. Professor Heinrich Friedrich Weber chaired the Institute for Mathematical and Technical Physics, while Professor Jean Pernet chaired the Institute for Experimental Physics (*CPAE* 1, ed. note, 60).

THE ACADEMIC TRACK

After fourteen years of prior schooling, students entered the Polytechnic at a level comparable to the junior year of an American college, and they were typically about the same age as an American junior. Section VI A prescribed no set curriculum. The professors worked out a course of study in consultation with each student. This included graded courses in their major fields of study along with electives within and outside their fields for which official grades were not recorded. After two years the students took the intermediate diploma exams in order to continue their studies toward the diploma. At the end of four years of study they took the final exams for the diploma, a professional certification comparable to a master's degree in the United States. Like a master's degree, then and now, it typically required eighteen years of education, a research thesis (graded in Europe), and one or more final oral exams. Moreover, like a master's degree, a diploma was required as certification for teaching and for many other professions.

The Polytechnic did not grant the doctoral degree until it became the Swiss Federal Institute of Technology (ETH) in 1911. Prior to that time Polytechnic physics students who wished to obtain a doctorate, as Marić and Einstein did, could take advantage of an arrangement whereby they completed their dissertation research under the direction of a Polytechnic physics professor; but the University of Zurich would actually confer the degree with the consent of its physics professor Alfred Kleiner, who served as the second referee (Stachel 2005, 34–35). Theoretical physics was then only just emerging as an independent discipline, primarily in Germany, from its roots in applied mathematics and mathematical physics (Jungnickel and McCormmach 1986, 287–297). Because experimental work dealt with verifiable data, it was considered at the time more fundamental than theoretical work. Although Einstein did not then appreciate the significance of mathematics for physics (Einstein 1979, 15), it was he who later brought theoretical physics to both the university and the Polytechnic as an independent discipline of equal status to other physics disciplines. During the late 1890s all physics research at the Polytechnic, and elsewhere, was thus primarily experimental in nature, although due attention was given to the supporting theories and mathematical foundations.

Polytechnic physics students pursuing a doctorate had to prepare a publishable dissertation demonstrating their expertise in experimental research. This usually took about a year. They did not, however, need to take any additional course work. Those who wished to teach and research at the university level generally had to receive, in addition to the doctorate, a higher certification, the Habilitation. This required submission of a high quality original research publication and an oral examination in the candidate's field conducted by the entire faculty. There was no tenure process as in US universities. Once a candidate was "habilitated," he or she was eligible to be "called," or appointed, permanently to any open full professorship (teaching chair) in the field by state authorities in consultation with the local faculty. Research and teaching at the university level were not considered a job but a calling, and professors, as highly paid civil servants, enjoyed high cultural and social status.

FIRST-YEAR COURSES

Entering the Polytechnic in fall 1896, Einstein, like Marić, found lodgings in student boarding houses. Their first-year courses were, as often today, heavily oriented toward mathematics. We are fortunate to have complete transcripts and exam grade reports for both Mileva and Albert throughout their higher education (see the appendices). The results varied widely for both, owing to ability and wavering interest. But in comparison with their fellow mathematics students, neither student gave evidence of mathematical brilliance. Walter Isaacson (2007, 52) described Mileva as "an intense but not a star intellectual."

An average of Mileva's first-year grades for the seven graded courses taken during both semesters (six in mathematics, one in Mechanics with Exercises) was about 4.29 of 6; her highest grade was 4.5, her lowest 3.5 (see appendix C). Einstein's first-year math grades for the same courses as Marić's (also listed in appendix C) were the same or slightly higher than hers, with two high grades of 5 and a low grade of 4, yielding an overall average for all seven courses of about 4.64. (Content descriptions of their courses are given in *CPAE* 1, appendix E.)

THE HEIDELBERG INTERLUDE

As classmates, Mileva and Albert frequently saw each other, and they probably discussed their work often that first academic year at the Polytechnic (1896–97), even perhaps during the summer hike they took together after classes ended. But for reasons unknown, Mileva left Zurich in fall 1897 for Heidelberg, Germany, where as a registered auditor she attended university lectures in physics and mathematics. The math and science faculty at Heidelberg had admitted several women in the early 1890s, but the general admission of women as full-time students did not occur until 1900 (Eckart 2001, 3–5;). Among Mileva's favorite professors there was the experimentalist Philipp Lenard, a future Nobel laureate for his later work on the photoelectric effect—the ejection of electrons from metals by light— as well as an infamous future Nazi who opposed Einstein's work as "Jewish physics" (Beyerchen 1977, chap. 5). After one semester she returned to Zurich to resume her full-time studies at the Polytechnic and submitted for her student record (Matrikel) an "attendance certificate from the University of Heidelberg" (T-G 1983, 56; 1988, 60; Marić, Student Record).

During Marić's Heidelberg interlude a noticeable change occurred in their platonic relationship: she and Einstein began their now famous correspondence, which continued into their marriage whenever they were apart, and sporadically thereafter. Albert appears to have initiated the correspondence with a remarkable four-page letter, now unfortunately lost. Mileva responded that fall, sometime after October 20. In this first letter she used the customary German formal pronoun *Sie* for the English "you":

> It's been quite a while since I received your letter, and I would have answered it immediately to thank you for your sacrifice in writing a four-page letter, thus repaying a bit of the enjoyment you gave me during our hike together—but you said I shouldn't write until I was bored—and I am very obedient (just ask Miss Bächtold). (Einstein and Marić 1992 [hereafter E-M], 3)

Mileva also told him in her letter that, having returned home in the meantime, she had informed her father about her classmate, a Mr. Einstein. Perhaps she also showed him the four-page letter. As a result, she continued:

> Papa gave me some tobacco that I'm to give to you personally. He's eager to whet your appetite for our little land of outlaws. I told him all about you—you absolutely must come back with me someday— the two of you would really have a lot to talk about! (E-M, 3)

Without further information, especially Einstein's four-page letter, we can only speculate about the reason for Mileva's sudden departure for Heidelberg. It was not uncommon for students at that time to visit another university for a semester, or even a year. Given her enjoyment of the summer hike and her apparently favorable report to her father about her classmate, the awakening of romantic feelings is also a possibility, beside the adventure of visiting Heidelberg. Charles S. Chiu assumes that interpretation. But he attributes the alleged romantic feelings mainly to Einstein, while interpreting Mileva's departure as an attempt to escape his amorous advances. "For a year," Chiu (2008, 33) writes, "Albert Einstein showers his 'little fugitive' with letters, in which he asks her, above all, to return to him."

We know of only three letters from Albert to Mileva during that single semester (October 1897–March 1898), of which only one survives, These are Albert's lost four-page letter that initiated the correspondence, an empty envelope addressed to her and postmarked January 2, 1898, and his surviving letter, tentatively dated by the Einstein editors as February 16, 1898. It opens with the standard formal salutation Esteemed Miss (*Geehrtes Fräulein*). Hardly in hot pursuit, Albert apologized for "not responding to your letter for such a long time." Then he added, apparently referring to a lost intervening letter from her in which she announced her decision to return to Zurich, "I'm glad that you intend to return here to continue your studies. Come back soon; I'm sure you won't regret your decision. ... You will, of course, have to give up your old pleasant room which a Zurich philistine now occupies ... serves you right, you little runaway [*Ausreisserin*]!" (E-M, 4–5, latter ellipses per original) The last remark seems more a joke than an accusation. Mileva managed to find another room in Miss Bächtold's rooming house. He then went on to summarize the course work she had missed, which, he says, he can explain more fully when she returns, probably by showing her his class notes. The notes for part of the course have survived and were published in the first volume of his *Collected Papers* (*CPAE* 1, doc. 37).

That Marić was not neglecting her studies in Heidelberg is indicated by a report in her first letter to Einstein, in which she talks about one of Lenard's lectures she had found particularly interesting:

> It really was too enjoyable in Prof. Lenard's lecture yesterday; now he's talking about the kinetic theory of gases. It seems that oxygen molecules travel at a speed of over 400 m per second, and after calculating and calculating, the good professor set up equations, differentiated, integrated, substituted, and finally showed that the molecules in question actually do move at such a velocity, but that they only travel the distance of 1/100 of a hair's breadth. (E-M, 4)

The kinetic theory of gases was one of the triumphs of nineteenth-century physics, and it is a topic found in the introductory courses of most physics programs, then and now. In his reply, Albert listed the kinetic theory of gases as one of the main topics included in Professor Weber's second-year lectures that she had missed (E-M, 5). He himself would go far beyond such elementary ideas in his own independent studies in advanced physics during the coming Polytechnic years (*CPAE* 2, ed. note, 41–55).

INDEPENDENT STUDIES

While Marić and Einstein's first year of study had consisted almost entirely of mathematical subjects, their remaining three years focused on physics. But Einstein was gradually growing restless, and Weber's courses were much to blame. During Mileva's sojourn in Heidelberg, Albert wrote to her that he too was impressed with the kinetic (or dynamic) theory as it was presented at the time in Weber's physics course. He received a 5.5 and a 5 for the two semesters. "Weber lectured masterfully on heat (temperature, heat quantities, thermal motion, dynamic theory of gases)," he wrote. "I eagerly anticipate every class of his" (E-M, 5). But he became disillusioned when he found that this and subsequent courses contained no physics topics closer to the present. In particular, James Clerk Maxwell's electrodynamics and his electromagnetic theory of light, foundations of modern physics that were of special interest to Einstein, remained painfully absent. "Anything after Helmholtz was simply ignored," his classmate Louis Kollros (1956, 29) recalled.

Einstein and Marić began to strike out on their own. Einstein (1956a, 10) recalled: "I followed several lectures with great interest. But otherwise I 'cut classes' often and studied at home the masters of theoretical physics with holy zeal," an enthusiasm he had already indicated in a note to Marić (written sometime after April 16) when both were back in Zurich for the start of the 1898 summer semester. He told her he had borrowed from her lodgings an extracurricular textbook by the theoretical physicist Paul Drude, which they were planning to study together (E-M, 6). It was probably Drude's 1894 text on electromagnetism, *Physics of the Ether on the Basis of Electromagnetism* (*CPAE* 1, doc. 40, note). Shortly thereafter he informed her that he had already finished half of the "tome," apparently while ill at home in his student room. He then extended to all of Miss Bächtold's boarders an invitation to a "get-together" in his lodgings, even though he was still not feeling well. "I hope you are among those who come," he wrote. "But if you can't make it, I'll visit you as soon as I feel well enough" (E-M, 6).

Mileva and Albert were fast becoming close friends and study partners. In another surviving note from 1898 (likely written after November 28), he told of the death of the brother of his close friend Michele Angelo Besso, a former Polytechnic engineering student whom Einstein, and probably also Marić, had met during an evening musical gathering in Zurich. He closed this note with a touch of sadness: "If you don't mind, I'd like to come over this evening to read with you. Your Albert Einstein" (E-M, 7).

INTERMEDIATE EXAMS

Summer 1898 marked the end of Mileva and Albert's second academic year at the Polytechnic. Their intermediate diploma exams were scheduled for October, just before the start of their third year. During 1897–98, Mileva had been absent the first semester, leaving a blank slate in her Polytechnic grade report (Marić, Student Record). Albert's record shows that, in addition to Weber's physics course, he enrolled that semester in another course on mechanics (grade 5.5), as well as in courses on differential equations (5) and projective geometry (4). When Marić returned for the second semester, they both registered only for Weber's physics course for a grade within their major field, and both received a 5 (appendix C). They filled in the rest of their schedule with ungraded electives within and outside their major subject area. Marić's outside electives included a course in psychology as well as those titled Geology of Mountains, Foundations of National Economy, Botanical Excursions, and Swiss Cultural History (Marić, Student Record; T-G, 1983, 57; 1988, 61). Einstein's outside electives included the Philosophy of Kant and Goethe, as well as such economic subjects as Bank and Stock Market, the Social Consequences of Free Competition, and Mathematical Foundations of Statistics and Personal Insurance (*CPAE* 1, doc 28). The last would have proved useful during his desperate consideration of a post-Polytechnic job in the insurance industry when an academic post eluded him (E-M, 25 and 53).

Because of the semester spent in Heidelberg, Marić was permitted to take the intermediate diploma examination after her third year, in early October 1899, rather than after her second year when Einstein and the math students in their class in Section VI A took it. Both passed. According to an "Addendum by the Editor," a Zurich city archivist who inserted it into the reprint editions of Trbuhović-Gjurić's book in 1988 and 1993 (Zimmermann 1988a, 63), Marić's intermediate exam results were generally good, ranging from 4.75 in Descriptive Geometry and Geometry of Position to 5.5 in physics (see appendix D). This resulted in a decent grade average of 5.05, but it nevertheless put her in fifth place among the six students who took the exams with her, most of whom were mathematics majors. The same grade average for exams taken the previous year, along with Einstein and their mathematics classmates, would have put her in last place (*CPAE* 1, doc. 42). It was the 4.75 in the dual subject Descriptive Geometry and Geometry of Position that brought down her average. In a letter to Einstein from her family's vacation home near Kać in late summer 1899, Marić wrote that, in addition to cultivating the family garden, she was "cramming" for the exams and complained that Descriptive Geometry "is the hardest to master" (*CPAE* 1, doc. 53; E-M, 12).

Einstein's performance on the intermediate exams turned out to be exceptional (appendix D). He received an overall 5.7 (of a possible 6), the highest average grade among those who took the exams with him (*CPAE* 1, doc 42). His grade of 5.5 in physics was the same as Marić's. But his 5.5 in Descriptive Geometry and Geometry of Position, compared with her 4.75, made the biggest difference, along with 6s in analytic geometry and mechanics to her 5s. Two years later his performance on the final exams would be quite different.

Although Einstein had done well on the intermediate mathematics exams, he tended to focus in his course work only on the topics that interested him and, as we know from an earlier quote, he "cut classes" that did not. Einstein also recollected attending ungraded mathematics courses offered by Carl Friedrich Geiser on differential geometry and by Hermann Minkowski on potential theory and the theory of functions. Einstein was especially "fascinated" by Geiser's course on differential geometry, a subject that not only foreshadowed but also proved essential to his future work on the general theory of relativity (Einstein 1956a, 10–11). But Einstein cut classes in Minkowski's course, and his teacher considered him "lazy" because of it. Although Minkowski would later play a key role in the mathematical expression of the special theory of relativity, Einstein the student had yet to appreciate the value higher mathematics would have for physics. As he wrote in his "Recollections" of 1955:

> Higher mathematics didn't interest me much in my years of studying. I wrongly assumed that this was such a wide area that one could easily waste one's energy in a far-off province. Also, I thought in my innocence that it was sufficient for the physicist to have clearly understood the elementary mathematical concepts and to have them ready for application while the rest consisted of unfruitful subtleties for the physicist—an error which I noticed only later with regret. My mathematical talent was apparently not sufficient to enable me to differentiate the central and fundamental concepts from those that were peripheral and unimportant. (Einstein 1956a, 11)

When reviewing for the final diploma examinations in 1900, Einstein relied on the meticulous class notes recorded by his generous friend Marcel Grossmann, a mathematics student in Section VI A who would continue to play vital roles in Einstein's future work and career (Einstein 1956a, 11).

Figure 2.1
Left to right: Marcel Grossmann, Albert Einstein, Gustav Geissler, and
Eugen Grossmann in the garden of the Grossmann home in Thalwil,
Switzerland, May 28, 1899. Courtesy of the Albert Einstein Archives ©
The Hebrew University of Jerusalem. Photo [500–969].

STUDYING TOGETHER

One of the two letters we have from Einstein to Marić during their third year at the Polytechnic reveals a deepening relationship. In the letter to her from his home in Milan, written on either March 13 or 20, 1899, during a semester break, he recalled a recent evening spent with her in the home of the Polytechnic history professor Alfred Stern: "For hours I sat next to you, my charming table partner. It was then revealed to me in harsh tints how closely knit our psychic and psychological lives are" (E-M, 7–8).

By summer 1899 the barrier of social convention between the two students was beginning to crumble. He began addressing her by the affectionate nickname "Dollie" (Doxerl, in Bavarian dialect) and signing his letters with his first name alone (though still using the formal *Sie*). Writing to Dollie on vacation in Kać while he was on vacation in early August with his family in the Alps near Zurich, he told her of reading a book by Helmholtz on "atmospheric movements," and of his increasing attachment to her. "When I read Helmholtz for the first time I could not—and still cannot—believe that I was doing so without you sitting next to me. I enjoy working together very much, and find it soothing and less boring" (E-M, 9).

For Marić and Einstein, the third year at the Polytechnic, 1898–99, focused on beginning laboratory work in preparation for their thesis research. Both enrolled during the first semester in Jean Pernet's Physics Practicum [experimental physics] for Beginners, and during the second semester in Weber's course Scientific Projects in Physics Laboratories. Einstein also enrolled in Weber's laboratory course on electro-technology, the content of which was likely familiar to him, at least in part, from his family's business. He received a grade of 6. By this time Marić and Einstein were becoming so engrossed in outside readings in theoretical physics that Einstein apparently regarded Pernet's lab course as dispensable. Pernet would have none of it. He gave Einstein a failing grade of 1 for the course and, as recorded on his official student record, Einstein received a "reprimand from the Director for [his] non-diligence" (*CPAE* 1, doc. 28). Marić received a 5 in the course. Both received 5s for Weber's second-semester lab course (appendix C).

In another letter from the Alps sent to Dollie in Kać, written most likely on August 10, the "non-diligent" Einstein revealed how far his independent theoretical physics studies had taken him. Having just finished a collection of papers by Heinrich Hertz (1892) on the propagation of the electromagnetic force— a topic not covered in his course work—he made an astonishing statement:

> I am convinced more and more that the electrodynamics of moving bodies as it is presented today doesn't correspond to reality, and that it will be possible to present it in a simpler way. The introduction of the term "ether" into theories of electricity has led to the conception of a medium whose motion can be described, without, I believe, being able to ascribe physical meaning to it. (E-M, 10)

Such notions at the time foreshadowed some of the ideas of his later special theory of relativity, which, after much further development, he presented in one of his three famous papers of 1905, titled "On the Electrodynamics of Moving Bodies" (*CPAE* 2, doc. 23).

In one of her rare surviving letters to Einstein from this period, written between August 11 and September 10, 1899, Marić replied from Kać to Einstein's two letters: "Both of your letters have found me contented at our country retreat; I thank you for them and look forward to receiving another one soon" (E-M, 12). Aside from her reference to "cramming" for the upcoming intermediate exams, cited earlier, she did not mention anything about science in this letter, and not a word about Einstein's new ideas on electrodynamics. But she did assure him that "your every letter gives me warm memories of home [Zurich]. Our series of shared experiences has secretly given me a strange feeling that is evoked at the slightest touch ... and makes me feel, so it seems, as if I were in my room once again" (E-M, 12). Given the context of Einstein's letters in an extended period separated from her, and her lack of response to his new ideas, it seems unlikely she was participating at that point in the formulation of these ideas.

CIRCLES OF FRIENDS

At the start of the summer semester in April 1899, Mileva moved from Miss Bächtold's house to Pension Engelbrecht, a Zurich boarding house run by Mrs. Johanna Engelbrecht at Platenstrasse 50. Five other female students from central Europe were in residence. Three would become Mileva's close female friends in Zurich: Ružia Dražić from Serbia (whom she had met earlier in Šabac); Milana Bota, a psychology student from Serbia whose family stemmed from Alsace, France; and Helene Kaufler (the daughter of a prominent lawyer and his wife, who was born to an aristocratic family), from Vienna. Helene was studying history at the University of Zurich and would become Mileva's closest friend and confidante. (The other two boarders were Miss Popova and Miss Ivanović.) In addition to knowing Ružia, Mileva had already been acquainted for at least a year with Milana and probably also with Helene. It was likely through them that she learned of a vacancy at the pension, where she could be with her friends. In 1898 Bota described Mileva to her mother as "a very good girl, clever and serious, she is small, frail, dark, ugly, talks like a real Novi Sad girl, limps a little bit, but has very nice manners" (quoted in M-KS, 5).

Mileva moved into her new room on April 18, 1899. The next day Helene Kaufler reported to her mother, "Since yesterday we've had a mathematician here, a Miss Marić from Serbia, but I have not seen her yet as she only lodges here but does not board" (M-KS, 53). On May 29 she reported: "This afternoon I first played [piano] four hands with Miss Bota, who is just as musical as Ma [mother], then I studied hard again, had tea with Miss Bota and Miss Marić, and then went to Mrs. Tarmau" (M-KS, 53).

Albert had also developed a circle of friends. It included his mathematics classmates Marcel Grossmann and Jakob Ehrat, as well as Michele Angelo Besso, the engineering graduate from the Polytechnic. Besso came from Trieste and Rome and was at that time working as an engineer in the nearby town of Winterthur. In 1898 he married Anna Winteler, the daughter of Jost Winteler who had earlier hosted Einstein at the Aargau Cantonal School (*CPAE* 1, 378).

Einstein was also a frequent visitor at Pension Engelbrecht even before Marić's move, sometimes joining her for musical evenings with her future housemates. Marić played the Serbian tamburitza (a long-necked lute) and Einstein the violin. Bota wrote to her parents in 1898, "I wanted to answer your letter yesterday afternoon, but I had a visit, it was Miss Marić with the German I told you about and we made music the whole afternoon. ... I wish you could hear him play—he plays with a deep passion and full sound. He is a real artist" (quoted in M-KS, 4).

By September 1899 Marić was busily studying in Kać for her intermediate exams the following month. Einstein wrote to her at the end of September from Milan to encourage her with a play on words regarding Professor Wilhelm Fiedler, the instructor of descriptive geometry whose name means "fiddler" in German. "The fiddling should be behind you by the time you get this," he wrote. "I've been thinking about it so much lately. I'm sure everything will go well—your hard little head assures me of that. If only I could look through the keyhole!" Later in the letter he expressed how much he looked forward to the enjoyment they would have in being back together again. "When I'm back in Zurich, the first thing we'll do is climb the Ütliberg [a nearby mountain overlooking Lake Zurich and the city]. Then we can take pleasure in unpacking our memories of the Säntis [highest mountain in northeast Switzerland]. I can already imagine the fun we'll have. And then we'll start in on Helmholtz's electromagnetic theory of light, which I still haven't read—(1) because I'm afraid to, and (2) because I didn't have it" (M-KS, 15–16).

DIPLOMA RESEARCH

Beyond the Alpine fun and Marić's intermediate exams, the two physics students prepared to face their final diploma exams at the end of the coming school year, 1899–1900. During that year they also needed to complete laboratory research projects for their diploma theses. Both enrolled again during their fourth year in Weber's Scientific Projects in Physics Laboratories, and both received the same good grades of 6 and 5 for the two semesters. Yet, for unknown reasons, the two candidates received the lowest grades for their thesis research projects of the five students taking the diploma exams. Marić received a 4, Einstein 4.5 (of 6) (*CPAE* 1, doc. 67). (Weber was not responsible for grading the other three candidates, who were mathematics students submitting mathematics theses.)

In Section VI A, each student could choose a thesis topic with the consent of the departmental professors' committee. Burdened with preparation for her intermediate exam, Marić had not yet settled on a topic by the start of the school year. Einstein, still pondering the problem of the electromagnetic ether, had written to her on September 10, 1899: "In Aarau I had a good idea for investigating the way in which a body's relative motion with respect to the luminiferous ether affects the velocity of the propagation of light in transparent bodies. I even came up with a theory about it that seems quite plausible to me. But enough of this! Your poor little head is already crammed full of other people's hobby horses that you've had to ride. I won't bother you with mine as well" (E-M, 14). A month later Marić was still cramming her head for the exams. Einstein's "good idea" and his plausible "theory" may be seen as further preliminary ideas related to the later special theory of relativity, but there was no indication yet in the letters that she was involved with his "hobby horse."

Weber rejected Einstein's proposed experiment, probably because it was similar to an earlier experiment on the hypothetical ether by the French scientist Armand Hippolyte Fizeau. Einstein then submitted a second proposal that he thought would appeal to Weber. It derived from one of Weber's specialties that had appeared in his second-year lecture notes (*CPAE* 1, doc. 37). He proposed an investigation of the relationship between the conductivities of heat and electricity in different materials in connection with the newly discovered (1897) electrons found to exist in all materials. "The procedures can be carried out very simply and require no equipment that is not readily available to us," he wrote Marić on October 10 (E-M, 18). But again, the professor rejected it, and again the reasons are unknown. Perhaps it was too simple, too theoretical, or perhaps too much a joint project with Marić. Finally, Mileva and Albert worked out together a two-part project on heat conduction that they presented to the professor, who finally concurred, to their great relief. Mileva's project dealt, in part, with the temperature dependence of specific heats of various materials (E-M, 30). She wrote from Zurich to "My dearest Miss Kaufler" on March 9, 1900: "Professor Weber has accepted my proposal for the diploma dissertation, and he was even quite pleased with it. I am looking forward to the research I shall have to do. E. has chosen for himself a very interesting topic" (M-KS, 60). In July 1900, Kaufler reported to her mother:

> Miss Marić and Mr. Einstein have now finished their written examinations. They devised their [thesis] topics together, but Mr. Einstein relinquished the nicer one to Miss Marić. He will most probably become assistant to his professor and remain here. Miss Marić was also offered an assistantship at the Polytechnic, but because of the students she did not wish to accept it, she would rather apply for an open position as librarian at the Polytechnic. (M-KS, 60–61)

There is no indication that Mileva applied for the library position and no clue why the students would discourage her from accepting the assistantship.

Neither the physics institute nor the Polytechnic administration preserved the diploma theses for Section VI A, so the reason for our students' low grades for their research, even when Weber gave them high marks in their parallel lab research course, remains a mystery. For Einstein the low grade could have derived from "non-diligence," as it did with Pernet. His research topic had little to do with what really interested him. He had chosen it solely to satisfy a requirement. Years later he recalled, "My and my first wife's Diploma theses concerned heat conduction and were for me without any interest" (quoted *CPAE* 1, doc. 63, n. 3).

DIPLOMA EXAMS

Marić and Einstein took the written and oral diploma exams during two days in late July 1900. The overall results for both were disappointing, owing mainly to their low thesis grades. The examiners obtained the total score for each student based on a weighted average, whereby the thesis grade was quadrupled; the subject grades for theoretical physics, practical (experimental) physics and theory of functions were doubled; and the astronomy grade was left as is (thus the grade was the total points divided by 11). Both students would have to do very well on the subject exams in order to counter the heavily weighted low thesis grade. For his subject grades, Einstein received 5 in theoretical physics, practical physics, and astronomy; and 5.5 in theory of functions (see appendix D). But his quadrupled thesis grade of 4.5 brought his overall score down to 4.91, which placed him fourth among the five students who remained in their group (one of the math students, Louis-Gustave du Pasquier, had dropped out). On July 28, 1900, Einstein was one of the four students in his class awarded a diploma that year (*CPAE* 1, doc. 67). He was now certified as a Specialized Teacher in Mathematical Instruction (*Fachlehrer in mathematischer Richtung*) (*CPAE* 1, 375).

Mileva was the unfortunate fifth student. She performed considerably less well on these exams than she had on the intermediate exams, with an especially poor showing of only 2.5 on the mathematics exam (theory of functions). Together with the 4 for her diploma thesis, the result was a disappointing overall score of only 4.00. The low average and probably the low math grade prompted the section head, on behalf of the Committee of Examiners (all of the section professors), to recommend that Fräulein Marić not be granted a diploma (*CPAE* 1, doc. 67). It is not known exactly how the committee reached this conclusion. No minimum passing grade or other performance requirements are stipulated in the records or grade reports (see documents on website with Marić, Student Record).

Mileva was devastated. Yet, as so often in the past, she was determined to succeed. She vowed to take the exams again the following year, and to work on her doctoral dissertation in the meantime. Einstein strongly supported her. "I'm so lucky to have found you, a creature who is my equal, and who is as strong and independent as I am! I feel alone with everyone else except you," he wrote on October 3, 1900 (E-M, 36).

COLLABORATION

As they prepared for their final exams in 1900, Mileva and Albert were growing ever closer. Earlier that year she and Albert were on a first-name basis and using the familiar pronoun *du,* reserved for close friends. And both now had nicknames for each other—Dollie (Doxerl) and Johnnie (Johonzel). In early 1900 she urgently sent Johnnie a note written with echoes of Bavarian dialect (in the original):

My Dear Johnnie,

Because I like you so much, and because you're so far away that I can't give you a little kiss, I'm writing this letter to ask if you like me as much as I do you? Answer me *immediately.*

A thousand kisses from your Dollie (E-M, 18)

By the time of their final exams Dollie and Johnnie had decided on marriage. In addition, they had become closer partners in the study of physics texts and even apparent collaborators on Einstein's original research at the time. In his letters to her, he seemed explicitly to acknowledge her role. Evan H. Walker (1991, 122) has found thirteen instances in the letters in which this occurred during the period from September 1900, following Mileva's first failure of the diploma exams, to December 1901, following her second failure. Einstein wrote, for example (emphases added): "If a law of nature emerges from this, *we* will send it to Wiedemann's Annalen" (October 3, 1900); "How happy and proud I will be when the two of us together will have brought *our work on the relative motion* to a victorious conclusion!" (March 27, 1901); and "*our theory* of molecular forces," which appeared twice (April 15 and December 12, 1901) (E-M, 35, 38, 44, 66).

From the above quotations, Walker (1991,123) concluded, "It would seem that Mileva Marić deserved to be a co-author, and her name should have appeared on the original 1905 [relativity] paper." Nearly every popular author since then has agreed, and they have extended this conclusion to his other 1905 papers (see part III). But, while accepting Marić's early role, Einstein scholars and historians have generally concluded that she could not have been a co-author of the relativity paper, which was written after four more years of complicated development, partly in close concert in the end with his friend Michele Besso, whom he acknowledged.(e.g., Martínez 2011, 194–195). Whether or not she discussed his relativity and other 1905 works with Einstein, provided practical support, or participated in any other way unfortunately cannot be answered definitively by the evidence available to us.

Returning to the early years, Einstein submitted his first paper for publication in December 1900. It appeared in Wiedemann's *Annalen der Physik* (Annals of Physics) on March 1, 1901, under the title "Conclusions Drawn from the Phenomenon of

Capillarity" (*CPAE* 2, doc. 1, Engl.). Einstein was listed as the sole author, yet on April 4, one month after publication, he told Marić of a recent meeting with Besso in which he had discussed with him a number of topics including molecular forces and surface phenomena. "He is very interested in *our investigations*, even though he often misses the overall picture because of petty considerations" (*CPAE* 1, Engl., doc. 96, emphasis added). On October 3, 1900, two months before submitting the capillarity paper for publication he wrote her:

> The results on capillarity, which I recently found in Zurich, seem to be totally new despite their simplicity. When we come to Zurich, we shall seek to get empirical material on the subject through Kleiner. If a law of nature emerges from this, we will send it to Wiedemann's *Annalen* [*der Physik*]. (*CPAE* 1, English, doc. 79)

We have no letters from Marić responding to Einstein's letters, nor does he mention receiving any letters from her on these subjects. Thus, we have no information at all on what exactly she contributed to the paper on capillarity. Yet it does seem from the above and other statements in his letters that she was indeed working with him, especially on the capillarity paper. Despite this, on December 20, shortly after he submitted the paper for publication, she attributed its authorship solely to Einstein in a letter to her close friend Helene:

> Albert wrote a paper on physics that will probably soon be published in *Annalen der Physik*. You can imagine how proud I am of my darling. This is not just an everyday kind of paper but is a very important one; it deals with the theory of liquids. (M-KS, 70)

This does not rule out her collaboration on this paper, but without knowing more details, such as why they decided not to include her as a co-author, we cannot know what and how extensive that collaboration might have been.

MARRIAGE PROSPECTS

Marić and Einstein were not alone among their friends in their decision to marry. By their final exams university students were typically in their mid-twenties, and parents and students both expected marriage, jobs, and babies to follow quickly (not necessarily in that order), as they did for Mileva's friends. Both Helene and Milana already had serious boyfriends, and both were married by the end of the year: Milana to Dr. Svetislav Stefanović, a Serbian physician, and Helene to Milivoje Savić, a Serbian chemical engineer working in Zurich on a stipend from a Serbian manufacturer. After their marriage the Savićs moved to Reutlingen in Württemberg, Germany, where Mr. Savić worked as a textile engineer. In 1901 Helene gave birth to her first daughter, Julka, and in 1903 a second daughter, Zora. Mileva's long-time friend Ružia Dražić died of tuberculosis in 1901 (M-KS, 64).

Despite Mileva's failure on the diploma exams in 1900, the prospect of marriage brightened the couple's hopes. But fate showed little regard for the optimistic Dollie and Johnnie. Matters grew worse on nearly every front. Albert had decided to inform his family immediately after passing the diploma exams of his plans to marry Mileva. In late July 1900, he traveled to Melchtal in the Swiss Alps to meet with his mother, sister, and an aunt for vacation. From there he wrote Mileva of the "scene" that ensued when he broke the news to his mother:

> So we arrive home, and I go into Mama's room (only the two of us). First I must tell her about the exam, and then she asks me quite innocently: "So, what will become of your Dollie now?" "My wife," I said just as innocently, prepared for the proper "scene" that immediately followed. Mama threw herself onto the bed, buried her head in the pillow, and wept like a child. After regaining her composure

she immediately shifted to a desperate attack: "You are ruining your future and destroying your opportunities." "No decent family will have her." "If she has a child you'll really be in a mess." With this last outburst, which was preceded by many others, I finally lost my patience. I vehemently denied that we had been living in sin and scolded her roundly.

On the next day, he wrote:

> Things were better. ... The only thing that is embarrassing for her is that we want to remain together always. Her attempts at changing my mind came to expressions such as: "Like you, she is a book—but you ought to have a wife." "By the time you're 30 she'll be an old witch," etc. (E-M, 19–20)

The mother's objections evidently centered on Mileva's age, her birth defect, and her career ambitions. In her view, he needed a domestic housewife to nurture and care for him while he pursued his career, apparently as she had done and as his second wife would do. The physical disability, then, was the primary problem, whereas differences in religious and national heritages did not appear of overt concern. A month or so later, returning home to Milan, Einstein wrote to his fiancée:

> Mama often cries bitterly and I don't have a single moment of peace here. My parents weep for me almost as if I had died. Again and again they complain that I have brought misfortune upon myself by my devotion to you, they think you aren't healthy. ... [His ellipses] Oh, Dollie, it's enough to drive one mad! You wouldn't believe how I suffer when I see how much they both love me. ... I'll only be able to recover from this vacation gradually, by being in your arms—there are worse things in life than exams. Now I know. (E-M, 29)

Mileva, too, suffered under the rejection of her "mother-in-law," as she called her in her letters to Helene. Einstein continued to comfort her with assurances of his enduring love and need for her and with dreams of their future life together. "How was I able to live alone before, my little everything?" he had written in August 1900. "Without you I lack self-confidence, passion for work, and enjoyment of life—in short, without you, my life is no life" (E-M, 26). That life would also include continued work with her. "No matter what happens," he assured her on September 19, "we'll have the most wonderful life in the world. Pleasant work and being together—and what's more, we now answer to no one, can stand on our own two feet, and enjoy our youth together. Who could have it any better?" (E-M, 33). Not until nearly two and a half painful years later, following Mr. Einstein's deathbed permission to marry and his subsequent death in October 1902, did Mileva and Albert finally marry.

Then there was the matter of supporting themselves. Family support greatly declined for both of them, owing mainly to their parents' financial problems. "It looks like my parents are broke again," Albert told Mileva in May 1901 (E-M 52; also, *CPAE* 8B, docs. 571, 572). Her father's income had declined after his early retirement for health reasons, while he had two other children to support at the university. Although increasingly independent of his parents, Albert did not want to marry Mileva against his parents' wishes and without having a job to support them both. Moreover, both had a year of doctoral research ahead of them in Weber's lab, and Mileva had to prepare to retake the diploma exams. Both of them had promises of paid assistantships with Weber before the exams, but nothing came of either offer. Having failed to obtain the diploma, Marić was ineligible for an assistantship, and Weber abruptly withdrew his offer to Einstein. The reasons are not known for certain, but Mileva wrote her friend Helene in December 1901: "You know that my darling has a very wicked tongue and on top of that he is a Jew.

From all this you can see that the two of us make a very sorry couple" (M-KS, 79).

THE WEBER CONFLICT

In addition to the falling out between Einstein and Weber, tensions were rising between Marić and Weber apparently over the quality of her doctoral research. Mileva had already planned to expand upon her thesis research for her doctoral dissertation, and Albert was again encouraging: "I'm also looking forward to working on our new papers. You must continue with your investigations," he wrote to her in September 1900. "How proud I will be to have a little PhD for a sweetheart, while I remain a completely ordinary person!" (E-M, 32). For his part, Einstein increasingly focused on the theoretical aspects of his dissertation research, which involved a study of intermolecular forces that led to his first two published papers (*CPAE* 2, docs. 1, 2). In this way, he could minimize his need to work in Weber's laboratory, but he would still need to confirm some of his results (*CPAE* 2, doc. 2, p. 39).

By winter 1901 Einstein was intending to enter the emerging discipline of theoretical physics as a career. It was a discipline founded mainly by European physicists, some of them hindered by anti-Semitism from gaining access to laboratories and professorships in experimental physics. This appears to have been Einstein's situation as well (Jungnickel and McCormmach 1986, 286–287). Mileva reported on his plan to Helene:

> Actually, we don't know what fate has in store for us. Albert applied for a practical job in Vienna, since he has to earn his living. He wants in addition to this job to continue improving himself in theoretical physics, in order to subsequently become a university professor. (M-KS, 72)

It is the strategy he followed, obtaining first a "practical" job in the Swiss patent office. But aside from her intention to obtain a doctorate in physics, she did not indicate her own career plans in research or teaching or in some related profession, or her strategy for achieving them.

In April 1901 Marić reentered the Polytechnic for the summer semester and, in anticipation of the final exams, she enrolled in the class on Determination of Geographic Location and in Weber's lab course Scientific Projects in Physics Laboratories for her doctoral research. She received a 5 in both courses (Marić, Student Record). Einstein, living practically hand to mouth, wrote her in May: "So, how is your work going, sweetheart? Everything going well in your life? Is old Weber behaving decently, or does he again have 'critical comments'?" (E-M, 51–52). Around that time, in May or June 1901, Mileva reported to Helene Savić that she had "several arguments with Weber, but I am used to it by now" (MKS, 76). The arguments appear to concern his "critical comments" regarding her dissertation, rather than something related to Einstein, who had left Zurich. If Weber had been already unfavorably disposed toward her or her research, as the low thesis grade of 4 implies, he would probably not have been very encouraging as she expanded that thesis into a dissertation.

Some have suggested that Marić's difficulty with Weber arose from his prejudiced attitude toward female science students; others have seen it as the result of Weber's anti-Semitic prejudice against Einstein, which brought Marić to Einstein's defense or which led Weber to attack his fiancée. That Weber harbored strong prejudices toward female students seems unlikely. As noted earlier, Section VI A had been welcoming female students since 1872, and the records indicate that Weber had previously

overseen other female research students without apparent difficulty (Stachel 2002, 33).

Although anti-Semitism may have been a motivating factor
in Weber's behavior, the evidence is very circumstantial. After
his attempts to obtain an assistantship at the Polytechnic failed,
apparently under opposition from Weber and Pernet, Einstein
began applying by letter for assistantships with other professors
in Germany and Austria. All were either rejected or went unan-
swered. He became convinced that Weber was sabotaging his
applications through poor letters of recommendation. "I would
have found such a job long ago, if Weber was not playing such
an underhanded game," he wrote his friend Marcel Grossmann
on April 14, 1901 (*CPAE* 1, doc 100).

At Mileva's urging, Einstein had been in touch with Weber
from Milan on March 23, to ask for his support and "to let him
know that he can't get away with doing such things behind my
back" (E-M, 36). It was to no avail. By then Einstein was
expanding his job search to Italy and to a former teacher in
Aarau. He was now avoiding German-speaking countries, which
included Switzerland, because, as he explained in a March 27
letter to Marić: "To begin with, one of the main obstacles in getting a position doesn't exist here [in Italy], namely anti-Semitism, which in German-speaking countries is as unpleasant as it
is a hindrance" (E-M, 39). It is not certain that anti-Semitism
was part of Weber's "underhanded game" or that he simply dis-
liked Einstein. But anti-Semitism in hiring for physics positions
was indeed prevalent in German-speaking countries, including
Switzerland (Jungnickel and McCormmach 1986, 286–287).
By now Einstein's friend Besso was in Milan and Trieste, and
through him he asked Besso's uncle, a professor, to make inqui-
ries about physics positions at Italian universities (E-M, 41–42).

RESEARCH PROGRESS

Einstein's research progressed nonetheless, and, when requested, he shared his progress with his fiancée. Responding in April 1901 to Marić's complaint that he was keeping his thoughts from her, Einstein wrote her: "Today I am going to give you a detailed report of what I'm up to, because I see that you enjoy it" (E-M, 42–43). He then proceeded to tell her about his thoughts on Max Planck's latest work on electromagnetic radiation in which Planck suggested what became the quantum of energy. It was a step toward Einstein's famous 1905 paper on the photoelectric effect and the light quantum hypothesis. Although other letters suggest her participation in his other work, she was now busy with her own work. There is no indication in this letter, written from his home in Milan, that she was participating in his thoughts at that time. Any reply from her has been lost, so we cannot know her response, if any, to his new ideas.

Einstein published his first paper, the one on capillarity, in March 1901 (*CPAE* 2, doc. 1). Even after distributing offprints of the paper to leading physicists following its publication, no job offers appeared, apparently owing to Weber's "underhanded game." Meanwhile he continued work on intermolecular forces for his doctoral dissertation. But rather than going through Weber, he decided to submit it directly to Alfred Kleiner at the University of Zurich in November 1901. "He will probably get his doctorate within a few months," Mileva wrote Helene in December. "I have read this work with great joy and real admiration for my little darling, who has such a clever head" (M-KS, 80). Kleiner rejected it—mainly because he did not like Einstein's "attack on the scientific establishment" that had failed to hire him (Isaacson 2007, 71). Kleiner told Einstein to withdraw the work voluntarily in order to obtain a refund of his

submission fee, which he did (*CPAE* 1, doc. 132). The dissertation has not survived, but some of the content appeared in his
subsequent published papers. He did not submit another dissertation to Kleiner until four years later, during the "miracle
year" 1905.

THE BATTLE OF DOLLIE

While Albert was traveling back and forth between Zurich and
Milan in order to fight what he called in October 1900 "the
battle of Dollie" (E-M, 35), Mileva's spirits fell. A new onslaught
against her from his parents sent Einstein to Milan and her into
despondency until his return. "I was already thinking that there
would be nothing more for me in this evil world," she wrote
Helene. "What really depressed me was that our separation
came about in an unnatural way, on the basis of slanders and
intrigues and the like." Albert's parents were trying to prevent
their marriage, she continued, "and you can imagine how I was
hurt to be attacked from that side." After Albert's return her
spirits suddenly improved, "I am happy that he loves me so
much," she wrote on December 30. "What more do I need?"
(M-KS, 69–70).

The lovers' misfortune seemed to break in April 1901 when
Einstein landed a temporary job as a substitute teacher from
May through July at a technical school in Winterthur, Switzerland, not far from Zurich. At the same time Marcel Grossmann
informed his friend that his father had recommended Einstein
to Friedrich Haller, the director of the Swiss patent office in
Bern, the capital of Switzerland. Haller and the senior Grossmann had been friends and colleagues for years (E-M, 44–45;
CPAE 1, doc. 100, and notes). With brighter job prospects and
an income that would allow for a brief vacation, Einstein invited

his fiancée in Zurich to meet him at Lake Como, in Italy, on his way from Milan to Winterthur. She accepted. But, as she explained on May 2, a letter from her parents, who were very doubtful of the young man whom they had yet to meet, "made me lose all desire, not only for having fun, but for life itself" (E-M, 48). Einstein's profession of love for her in his letter finally overcame her parents' doubts and her hesitation at being alone with him at a resort. "So I think we'll take that little trip after all," she wrote him on May 3 (E-M, 49).

A heavy snow was still on the ground, making a beautiful winter landscape, when Mileva met a certain "someone" in Como on May 5. That evening they hired a driver with a small horse-drawn sleigh having "just enough room for two people in love," and they rode up through the silent snow into the Splügen Pass through the Alps that separated Italy from Switzerland. All during the trip "I held my sweetheart tightly in my arms under the coats and shawls with which we were covered," she wrote Helene. The descent back down "was also beautiful," and then they tramped back to their lodgings through the heavy snow. "We had so much fun that we did not feel it burdensome at all" (M-KS, 75). "My God, how beautiful the world will look when I'm your little wife," she wrote Albert after returning to Zurich (E-M, 52). By the end of May 1901 his mother's worst fear was realized. Mileva discovered she was pregnant.

WITH CHILD

At this point, at age twenty-two, Einstein was on the verge of a career whose future, he might imagine, would be adversely affected by an unwanted child. More immediately, his teaching and patent office jobs as a public servant would be jeopardized if her scandalous pregnancy became known. He could have chosen the easy way out, as have many other men in a similar situation, but he did not. Instead, he wrote to Mileva at the end of May after receiving the news, "Be happy and don't fret, darling. I won't leave you and will bring everything to a happy conclusion" (E-M, 54). Yet the matter grew complicated for Mileva. Albert still would not marry her until he had a full-time job to support her and their child. "You just have to be patient!" he told her in the same letter. "You'll see that my arms aren't so bad to rest in, even if things are beginning a little awkwardly." He was more explicit in July. "As soon as I have such a position I'll marry you and have you come to live with me without writing a word of it to anyone until everything's been settled. And then no one can cast a stone upon your dear head" (E-M, 57). Until they were married, he said, it was better that she did not appear in Winterthur, the location of his teaching post. Instead, she remained in a room in Zurich where Albert visited her on Sundays, probably his only day off. When the Winterthur job ended in July, he began a long-term, part-time position in the town of Schaffhausen in September, where he tutored a wealthy English boy at a private boarding school; Mileva moved from Zurich to a nearby town, where again he saw her only on Sundays.

While Einstein's research on relative motion continued to progress, he insisted from Winterthur in early July 1901 that he was now willing to take any job available in order to marry her "*immediately*" (his emphasis). "My scientific goals and personal vanity will not prevent me from accepting the most subordinate role," he assured her (E-M, 57). It was probably of some help, for that same month Mileva, pregnant and unmarried, was scheduled to take her diploma exams for the second time. But then Albert was missing in action, first in Winterthur until around mid-July, then on vacation again with his mother in the Alps while Mileva took her exams alone in Zurich. "And now I wish you the best of luck on your exams and hope that they're over soon, dear sweetheart," he wrote from the vacation resort later in July (E-M, 59). He did not inform his parents of the pregnancy. Nor is there any indication that they informed Professor Weber.

MORE BAD NEWS

Marić retook the diploma exams in July 1901. She was now about two months pregnant, under great stress and probably wondering self-consciously if Weber was aware of her condition during the oral exams. Again, her thesis grade was 4, which was quadrupled to 16. It was probably the grade carried over from the previous year. If so, the odds were clearly stacked against her. Amid the distress of her personal life, she would again have to perform extremely well on the subject exams in order to balance out the heavily weighted thesis grade to yield a higher average than before. She did better on two of the exams, but worse on two others (see appendix D) The results were again an overall average of 4.00, which once again prompted the Committee of Examiners, headed by the mathematics professor Hermann Minkowski, to recommend that Fräulein Marić not be granted the diploma. Without a diploma, she still could not get a teaching job, and with Weber still the main physics professor, it would be difficult for her to continue at the Polytechnic toward a doctorate and a third try at passing the exams. "I have finished my studies," she wrote to Helene in fall 1901, "although thanks to Weber's concerns, I have not yet managed to obtain a doctorate. I have put up with a lot from him and will on no account go back to him again" (M-KS, 78).

Mileva now faced the prospect of returning to her parents with the news of her academic failure, in addition to her pregnancy (if they did not already know of it). She had written to Albert on July 8, before the exams, asking him to return with her to Kać: "My parents are probably in a better mood by now. Wouldn't you like to come along? It would make me so happy! ... And when my parents see the two of us together in front of them, all of their doubts will disappear" (E-M, 58). Albert declined. Several weeks later she asked him at least to write a letter to her father, presumably regarding Albert's plans to marry her. "Just write a short letter to my Papa; by and by I'll give him the necessary information, the unpleasant news as well" (E-M, 60). If he wrote such a letter, it has not survived. Mileva sadly returned home to face her family alone. After so many years of encouragement and support for their daughter as she struggled to overcome obstacles toward an academic career in science, their disappointment must have been palpable, surely as much as poor Mileva's, pregnant and alone at in her parents' home with no marriage in sight. Einstein did not accompany her home to meet her family for the first time until four years later.

Statistics show that out-of-wedlock births were not uncommon at that time in southern Hungary (Overbye 2001, 77–78). Nevertheless, that fact brought no comfort to Mileva, who was facing her family alone and desperately missing her fiancé. "My darling is far away from me again and it is so hard, so hard for me," she told Helene in fall 1901, "it makes my life bitter. ... Oh, Helene, pray to St. Peter for me that I might have him completely, that I do not have to be parted from him all the time—I love him so frightfully" (M-KS, 77). Mileva returned briefly to Switzerland, but then went back to Novi Sad in late November to await the birth of her child.

Already struggling and in a fragile emotional state, Mileva suffered another terrible blow, this time from Albert's parents, who sent a very nasty letter to her parents in December 1901. In the letter, she told Helene: "They reviled me so much that it was shameful. ... Both Albert and I suffered a great deal from it" (M-KS, 78). The letter could have been prompted by their fear that Albert would secretly marry Mileva without their permission. But given the vehemence of the letter, Helene's grandson Milan Popović, the editor of Mileva's letters to her, may be right: "the topic of this abusive letter was undoubtedly news of Mileva's pregnancy" (M-KS, 10). Mileva attributed authorship of the letter to her "dear mother-in-law." "This lady seems to have set as her life's goal to embitter as much as possible not only my life but also that of her son" (M-KS, 78). Her own parents seem to have been supportive.

During the months before the birth, as Einstein continued to tutor the English boy in Schaffhausen while waiting for the full-time job in Bern, the couple began thinking about the child's future. Already they playfully referred to the child in their letters as Lieserl, the Bavarian diminutive of Lise, reflecting Mileva's hope that it was a girl. As the birth approached, Mileva considered turning for support to Helene, who had just given birth to her first child. But she cautioned Albert somewhat cryptically on November 13: "I don't think we should say anything [to Helene] about Lieserl just now; but you should write her a few words every now and then. We must treat her well because she can help us with something important" (E-M, 63–64).

LIESERL

In contrast to Mileva, who was then suffering under the burden of Albert's parents, the forthcoming birth, the exam failure, and the end of her career ambitions, Albert was very upbeat in his letters to her in December 1901. His work was going well. He had just confidently submitted his doctoral thesis to Kleiner on

intermolecular forces, and he was busy at work on the electrodynamics of moving bodies, which, as he wrote on December 17,
"promises to be quite a capital piece of work" (E-M, 69). He was
overly optimistic: the capital piece of work had a long way to go,
through four more years of difficult development, before it became the special theory of relativity.

In addition to the above prospects, Albert was certain the job
at the Bern patent office would open soon, enabling them finally
to be together again, this time as a married couple. Such anticipation awakened dreams of recapturing the faded joy and free-
dom of their youthful student days—together with their Lieserl.
On December 12 he wrote:

> Just take good care of yourself and keep your spirits up and be
> happy about our dear Lieserl, whom I secretly (so Dollie doesn't
> notice) prefer to imagine a Hanserl. ... In two months' time [after
> getting the patent office job] we could find our lives brilliantly
> changed for the better, and the struggle would be over. I'm dizzy
> with joy when I think about it. I'm even happier for you than for
> myself. Together we'd surely be the happiest people on earth. We'll
> be students (*horrible dictu*) as long as we live and won't give a damn
> about the world. (E-M, 66, 68)

Albert even sought assurance from her that their revived student life would also include working together again, as before:
"When you're my dear little wife we'll diligently work on science
together so we don't become old philistines, right?" he wrote on
December 28 (E-M, 72–73).

Around February 1, 1902, Mileva gave birth in Novi Sad.
She received her wish for a daughter. Albert seemed pleased and
had earlier expressed his desire to keep the child with them.
"The only problem that still needs to be resolved," he had written in his December 12 letter, "is how to keep our Lieserl with
us; I wouldn't want to have to give her up. Ask your Papa, he's an

experienced man, and knows the world better than your over-
worked, impractical Johnnie" (E-M, 68). Her papa probably did
have a plan, but the details of Lieserl's fate are lost to history. The
rich insights provided by the Marić-Einstein and Marić–Kaufler
Savić letters in this period are silenced by critical gaps in the
extant letters. For the Marić-Einstein letters, the gap extends
from June 1902 to August 1903; during which the couple were
often together. For the Marić–Kaufler Savić letters—which are
all from Marić—the gap is from the end of 1901 to a dateable
letter in March 1903, then a further gap until May 1904—also
during the crucial period after Lieserl's birth. Walter Isaacson
states that the Savić family intentionally burned the Marić–
Kaufler Savić letters that presumably concerned Lieserl, in order
to cover up the child's existence (Isaacson 2007, 87).

THE LIESERL MYSTERY

Lieserl's existence and fate were so well hidden that she did not
become known until after the discovery of the Marić-Einstein
letters in 1986. By then, if she were still alive, she would have
been eighty-four years old. Despite extensive searches in Europe
and elsewhere, no one has yet discovered a trace of her or of
what became of her. Not even her registered name is known.
The official records were wiped clean, probably through Mr.
Marić's legal connections. No rumor or later memory from anyone involved ever slipped into public discourse—a remarkable
feat given the numerous hearsay reports eagerly repeated in later
semi-historical accounts of Mileva and Albert. It is a clear indication of how deeply the secret was buried.

Michele Zackheim (2000) and others have followed Lieserl's
trail as far as possible. Despite a few guideposts, most of the
story remains essentially pure speculation. It is of value here

only for what it suggests about the emotional impact on Lieserl's parents and their relationship. What we do know is that about five months after giving birth, sometime around the end of June 1902, Mileva returned alone to Switzerland without Lieserl. In a letter to Mileva dated by the Einstein editors as June 28, 1902, or later—his last before the gap—Albert wrote Mileva from Bern that he looked forward to seeing her on the weekend, indicating that she was somewhere nearby (E-M, 77).

Lieserl's birth around February 1 had been so difficult, that her father had to write Albert to tell him the news. "I was frightened out of my wits when your father's letter came," he wrote on February 4, 1920 (E-M, 73). But, rather than scurrying off to Novi Sad to comfort his fiancée and to see his newborn daughter, Einstein instead moved abruptly to an apartment in Bern in anticipation of the patent office job (E-M, 74). In so doing, he left his English student, the school director, and poor Mileva and the child in the lurch. According to Einstein's letter of February 17, the Swiss patent director Haller was about to advertise an opening for a patent examiner that same month in a newspaper advertisement tailored to Einstein's qualifications (E-M 76). Einstein duly applied, and on June 16, 1902, the Swiss Council appointed Einstein to the position of patent clerk, third class, with a salary of 3,500 Swiss francs, more than that of an associate professor at that time (*CPAE* 1, doc. 141). But marriage was still not forthcoming. For six months Mileva held out alone in separate lodgings in Bern and Zurich, traveling at times back to Novi Sad, until she and Albert were finally married (Hentschel and Grasshoff 2005). Shared lodging by an unmarried couple was a dismissible offense for a Swiss civil servant. In the end, Einstein probably never even saw his daughter.

The above sequence of events suggests that, even though Mileva and Albert had wanted to keep the child in December, by the time of her birth they had decided, with her father's advice or insistence, to give her up for adoption. Otherwise, Albert might have been there for the birth and Mileva would have stayed in Novi Sad with Lieserl until Albert was ready to marry and be reunited with them in Bern. In July 1901, after Albert first learned of her pregnancy, he had promised to marry her without telling anyone as soon as he had a job (E-M, 57). But now, gainfully employed and with Lieserl to be given up for adoption, he still put off the marriage until January 1903. (His father had finally granted him permission shortly before he died in October 1902.) It is conceivable that Albert now harbored reservations about the marriage and proceeded more out of duty than out of love. If so, he did not take the opportunity to abandon the distraught Mileva in those difficult times. But duty rather than love is usually not a sound basis for a lasting marriage.

One can hardly imagine the impact all this must have had on Mileva. Already emotionally battered, she must have been in great turmoil during this entire period. Suffering alone without her fiancé through a difficult labor, leaving a newborn baby after just five months, then living alone for six more months waiting for her fiancé finally to set a date for the marriage, must have been unimaginably devastating. No wonder her letters to Helene have disappeared. But there may have been another reason. Sometime in 1902 the Savić family moved from Germany to the vicinity of Belgrade, Serbia, with their one-year-old daughter Julka. Mr. Savić had accepted a job in the Serbian Ministry of Commerce and Industry (*CPAE* 1, p. 386). Helene's second daughter, Zora, was born there sometime in 1903. When she returned to Switzerland, Mileva had left Lieserl at home in the temporary care of someone until she could be placed in a permanent home. Although that someone was probably not Helene, who already had her hands full, Helene surely knew about Lieserl's care and discussed it with Mileva in their now-lost letters.

Mileva and Albert finally married on January 6, 1903, in a small civil ceremony at the Bern registry office (*CPAE* 5, doc. 4). It was Christmas eve on the Serbian Orthodox calendar. No family members from either side attended. The witnesses were two of Einstein's new friends in Bern: Conrad Habicht, a former math student at the Zurich Polytechnic; and Maurice Solovine, a Romanian philosophy student at the University of Bern, who had seen an advertisement Einstein had placed for tutoring in the Bern newspaper. The four of them celebrated over dinner that evening, and the newlyweds moved into a Bern apartment at Tillierstrasse 18 (*CPAE* 5, doc. 4, n. 1).

Dollie and Johnnie were together at last. But there was still one matter to settle. Eight months after her marriage, Mileva returned home for a month. It was apparently for the purpose of settling Lieserl's future. With Mr. Marić's help this would mean finally releasing her for adoption. But a difficulty arose: Lieserl had contracted scarlet fever, requiring Mileva's prolonged stay at home and prompting a complaint from Albert in a letter to her from Bern dated by the Einstein editors as September 19, 1903 (E-M, 78–79). She may have gone there initially because she had heard of Lieserl's illness. But that seems unlikely given Einstein's remark in apparent response to a report from Marić after her arrival: "I'm very sorry about what has befallen Lieserl. It's so easy to suffer lasting effects from scarlet fever. If only this will pass" (E-M, 78–79). If it did not pass, and she eventually died, that would explain what happened to her. But she was still alive at that time and adoption was still the plan, for Albert asked in the September 19 letter, "As what is the child registered? We must take precautions that problems don't arise for her later." As with most of Mileva's letters to Albert, her reply did not survive.

According to one post-1986 hearsay story, recounted by
Popović, Lieserl was baptized at the Kovilj monastery near Novi
Sad in September 1903, but then she died of scarlet fever shortly
thereafter (M-KS, 11). Other suggested possibilities (though
without clear supporting evidence) are that, disabled by scarlet
fever, she remained in the monastery (or more likely a related
convent) for the rest of her life; or that she was in fact adopted,
given a different name, and, with Mr. Marić's help, all traces of
her were expunged from the records.[1]

We don't know for sure what happened to Lieserl. Nor can
we hardly imagine how difficult it must have been for Mileva
never to see her baby daughter again. Most likely, she bore the
grief over the loss of her child alone all her life—the guilt and
self-doubt, the intense sorrow of separation. In most families
then, when a premarital child died or was given up for adoption,
no further word was spoken, no acknowledgment ever made
that the child had ever existed. Mileva's pain was buried deep.
Writing more than twenty years later, in summer 1924, she
reminded her friend Helene of "my unfilled desire for a daughter" (M-KS, 135). Mileva's loss, together with the loss of her
career, surely contributed greatly to her tendency to melancholy
thereafter and to its effects on her marriage and her intellectual
relationship with her husband. In the years ahead she would
need her husband even more, as he needed her even less.

3   AN UNSUCCESSFUL MARRIAGE

"I am now a married husband and lead with my wife a very nice comfortable life," Einstein reported to his longtime friend Michele Angelo Besso two weeks after his marriage to Marić. And, he was happy to report, "She takes excellent care of everything, cooks well and is always cheerful" (*CPAE* 5, doc. 5). Shortly thereafter, writing in March 1903 to Helene Kaufler Savić, Mileva had a different take on the early days of their marriage. "We have a nice little household, which I am taking care of quite alone, so you can imagine that, at least in the beginning, until I got used to it, I did not have much leisure time" (M-KS, 83). From the very beginning of their marriage, the traditional husband-wife division of labor between job and housework implied an apparently tacit agreement: he would have the main career, and any career she might possibly pursue would be secondary to her role as housewife. This must have been difficult indeed for her.

Einstein, too, was having difficulty with his work at the patent office. With eight hours per day, six days per week, plus private tutoring and his science, "I have terribly much to do," he wrote his friend Hans Wohlwend in 1902. But still, he told Wohlwend, he enjoyed the office work, "because it is uncommonly diversified and gives much to think about" (*CPAE* 5, doc. 2). In 1903, home alone after their marriage, Mileva was jealous of the time he spent away from her. The feeling would continue to grow. "I am even closer to my sweetheart, if it is at all possible, than I was in our Zurich days," she wrote Helene in March. "I am often angry at the boring office that takes away so much of his time" (M-KS, 83). Unbeknownst to his boss, Einstein had quietly resumed his search for an academic job.

DIVERGENT ROLES

The different roles that Einstein and Marić had assumed pulled them in different directions from the start. In January 1903, the month of his marriage, Einstein published the second of his three early papers on the fundamentals of statistical physics and the kinetic theory of molecules and atoms that formed the backdrop of his famous works of 1905 (*CPAE* 1, docs. 3, 4, 5). Aside from evenings and Sundays, Einstein managed at times to work discreetly on his research while at the patent office without his supervisor's permission, hiding his papers in a desk drawer.

After moving to Bern in February 1902, Einstein's new male companions were diverting his intellectual interests increasingly away from his wife, and both allowed this to occur. Einstein and his friends Conrad Habicht and Maurice Solovine so enjoyed discussing intellectual topics together with Einstein that they formed a regular discussion group, which they playfully called the Olympia Academy. Marić occasionally attended the "academy" meetings, but according to Solovine (1956, xii), "Mileva, intelligent and reserved, listened attentively to us, but never intervened in our discussions." This only reinforced her outsider status and her husband's intellectual bonding with his male friends rather than with her.

Soon after her marriage, Mileva seems to have given up entirely on a career. Although she could not have pursued an academic career without a degree, she had completed the four-year physics program at the Polytechnic for which she had received a leaving certificate showing courses and grades, but not the diploma. That might have enabled her to pursue a related career, perhaps as a librarian at the University of Bern or as a tutor. She might also have taken courses at the university toward another career. Instead, as far as we can discern, she seems to have settled for the role of domestic partner and provider of practical support and encouragement to her husband in his work. Indeed, even as his demanding work and career and his collaborations with others pulled him away from her, she frequently expressed pride in his success and may have identified with it. Yet still, it must have been for her a very uneasy arrangement, especially considering her sole responsibility for their later children, as her expressions of mounting frustration and unhappiness over the coming years suggest.

In August 1903, eight months after her marriage, Mileva traveled to her family in Novi Sad, a trip, discussed earlier, that likely concerned decisions regarding Lieserl. She stopped in Budapest, probably to change trains, and from there she reported to Albert on a postcard dated August 27. "It's going quickly, but badly. I'm not feeling well at all" (E-M, 78). She discovered she was pregnant again, and she worried that Albert would not be pleased. On the contrary, he wrote in mid-September, he was glad for her that she would have "a new Lieserl" (E-M, 78). The child was, instead, a Hanserl. On May 14, 1904, she gave birth in Bern to Hans Albert, whom they often called Albert, or little Albert. Mileva was overjoyed "with my dear little sweetheart" and, she was pleased to report to Helene on June 14: "His father is very proud of him and is already accustomed to behaving with fatherly dignity" (M-KS, 86).

As little Albert grew, big Albert enjoyed making toys and playing with him for long hours. He seemed to enliven their

marriage and bring them closer together again just as Einstein
headed into the intensive work of his "miracle year," 1905. But
also just in time for the miracle year, on Einstein's recommendation, the Bern patent office hired his friend Besso, who had earlier studied mathematics and physics at the University of Rome
and mechanical engineering at the Polytechnic. Besso remained
at the patent office until 1908 and quickly replaced everyone
else as Einstein's closest intellectual partner in this period.

THE YEAR OF MIRACLES

As Hans Albert's first birthday approached in 1905, a flurry of
Einstein's now famous papers began arriving almost monthly at
the *Annalen der Physik* and one at the University of Zurich.
These included (as cited with relevant doc. nos. in *CPAE* 2): on
March 18 his paper on the quantum hypothesis (doc. 14); on
April 30 his second doctoral dissertation submitted to the university (doc. 15); on May 11 his paper conclusively supporting
the atomic hypothesis in an explanation of so-called Brownian
motion (doc. 16); on June 30 the special theory of relativity
(doc. 23); and on September 27 his paper introducing the
equivalence of mass and energy (doc. 24). In addition he published numerous summaries of articles by others. It is plausible
that Marić helped in practical ways with this extraordinary outpouring of papers, such as proofreading and library research, but
we have no reliable evidence that she did, and no verified proof
that the woman who was trained in physics and mathematics,
and who was Einstein's wife, confidante, and supporter, contributed scientifically to these papers. Although some have argued
that Einstein required her assistance with the supposedly difficult mathematics, in fact the mathematics in the 1905 papers
(other than the dissertation) was not at all a challenge but actu-

ally basic for any student or graduate of the Polytechnic's program for physics and mathematics teachers.

More significantly, Einstein's 1905 *Annalen* papers sprang from deep within his nearly lifelong studies until then regarding such fundamental notions as atoms and molecular forces, electromagnetism and relative motion, and the nature of light and the ether. In his papers he brought the methods of statistical mechanics and scientific insight to bear on three different parts of physics:

1. the connection between the motions of invisible atoms and the motions of visible matter (Brownian motion);
2. the connection between the seemingly continuous light and the discontinuous ejection of electrons in the photoelectric effect;
3. and of course, the connections between space and time and between matter and energy.

No one else at that time was working to bring together such wide-ranging and fundamental ideas. Einstein's 1905 papers brought about a profound transformation in twentieth-century physical science and beyond through the establishment of three fundamental approaches to understanding the physical world deriving from the above: atomic theory, quantum theory, and relativity theory. And, amazingly, the mathematical expression of these new concepts required little more than the basic mathematics of a college physics student. It was truly a "year of miracles."

Although Einstein may have discussed his ideas with others, including his wife, only Besso received any acknowledgment at all in Einstein's 1905 papers, and it was given only in the relativity paper for which he had helped with some important insights (see Martínez 2009). "In conclusion," Einstein wrote, "I remark that in the work on the problems discussed here, my friend and colleague M. Besso stood faithfully at my side, and that I am indebted to him for several valuable suggestions (*manche wertvolle Anregung*)" (*CPAE* 2, doc. 23, p. 306 and Engl.).

In the midst of this amazing productivity, during his annual two-week vacation from the patent office in August 1905, Einstein finally traveled with Mileva and their son to Novi Sad, where she would introduce both big and little Albert to her entire family for the first time. Early the next year the University of Zurich awarded Einstein his doctoral degree (*CPAE* 5, p. 37).

THE CAREER LADDER

As the impact of the 1905 papers began to set in among Einstein's colleagues, the physics world began to take notice of the Swiss patent clerk, third class, who had produced them. Max Planck, the de facto "dean of German physics," dispatched his assistant Max Laue (later Max von Laue) from Berlin to Bern to inquire about the virtually unknown clerk. His visit became the beginning of a lifelong friendship and the first step toward eventually bringing Einstein to Berlin, the capital of German physics.

Einstein himself was already taking further steps toward an academic career at the university level. In 1907 he applied for "habilitation" at the University of Bern on the basis of his 1905 papers and prior work. With the additional certification beyond a doctorate he would be qualified to teach at the university level and eligible for permanent appointment as a professor at any university. But even with the fundamental publications of 1905, Bern University refused to confer habilitation without the formal submission of an original habilitation thesis. He finally acceded to their wishes under encouragement from Professor Alfred Kleiner, who saw him as a possible candidate for the University of Zurich. Einstein duly submitted a new paper on quantum physics, received the habilitation in February 1908, and began teaching part time, as was typical, at the University of Bern as a private lecturer (*Privatdozent*) (*CPAE* 5, doc. 80). The voluntary fees he received from the students were too meager for him to quit the patent office. In the following year, Einstein's sister Maria (Maja) received her doctorate at the same university. In 1910 she married Paul Winteler, the brother of Besso's wife.

Einstein's successful habilitation prompted Kleiner, who noted Einstein's mounting papers, to petition the University of Zurich for the creation of a new "extraordinary" (associate) professorship in theoretical physics, the first in this field in Zurich. But Einstein was not at the top of the candidate list. Friedrich Adler, the son of the leader of the Austrian Social Democratic Party, was ahead of him. But for various reasons the choice finally fell on Einstein for the job—that is, until Kleiner observed one of Einstein's lectures in Bern. Only after Einstein managed to improve his style and present a well-received lecture in Zurich did Kleiner submit a request to the faculty for approval of Einstein. And only after Kleiner managed to overcome anti-Semitic opposition within the faculty was Einstein finally appointed Extraordinary Professor of Theoretical Physics at the University of Zurich (Isaacson 2007, 149–153).

On October 15, 1909, Einstein began offering his first lectures as a professor. Just months before, the University of Geneva had awarded Einstein an honorary doctorate, and in September 1909 he had presented a now-famous lecture to the prestigious Society of German Scientists and Physicians meeting in Salzburg, Austria, in which he made the first strong argument for the so-called wave-particle duality in quantum physics: light appears to behave as both particles (quanta) and electromagnetic waves (*CPAE* 2, doc. 60). Unknown to Einstein, in 1910 the German physical chemist Wilhelm Ostwald nominated

Einstein for the first time as a candidate for the Nobel Prize in Physics for his formulation of relativity theory (Nobel Prize, Nomination Archive).

Einstein's career was on a rapid ascent, but the situation for Marić was heading rapidly downhill. Einstein's growing success and her growing frustrations, his devotion to his work and her need for attention and emotional support, his many friends and her isolation were all increasingly pulling them apart. Prior to moving to Zurich in September 1909, Mileva had informed her friend Helene of her husband's new position with a touch of sarcasm: "I cannot tell you how happy we are because of this change, which will free Albert of his daily eight hours in the office, and he will now be able to devote himself to his beloved science, and *only* science" (M-KS, 94, her emphasis). Preparing for the move, she wrote on September 3: "We are leaving Bern where I have lived for seven years now, lived so many lovely but I must also say bitter and hard days. ... I only hope and wish that fame does not have a harmful effect on his humanity" (M-KS, 98).

A month after arriving back in their beloved Zurich, Mileva discovered that she was again pregnant. On July 28, 1910, she gave birth to a second son, Eduard. The birth was a difficult one, and she was ill for weeks thereafter. The doctor, attributing the prolonged illness to overwork, recommended a live-in maid, which they could not afford on an associate professor's salary. Instead, her mother traveled from Novi Sad and stayed for a while to lend a hand (Isaacson 2007, 161).

During his first semester as a Zurich professor in winter 1909–10, Einstein offered a seminar and two lecture courses on the standard first-year subjects of mechanics (physics of motion) and thermodynamics (study of heat), in addition to his research (*CPAE* 11, p. 188). As the semester wore on, Einstein and Marić grew even further apart, and she, expecting Eduard, grew even more resentful of the time he spent devoted to work or distanced from his family. Referring again to Salzburg, she wrote Helene: "You see, with that kind of fame he does not have much time left for his wife. I read between the lines [of Helene's letter] a certain impish tone that I must be jealous of science. But what can you do? One gets the pearl, another the box" (M-KS, 102; a better translation for the original German—"*was kann man machen, dem einen die Perle dem andern die Truhe*" per Popović 1998, 221—is "To one the pearl, to the other the jewelry box.") Whatever she meant exactly by this, she saw herself on the losing end. Thanking Helene for her support and asking for her continuing love, she wrote during the winter semester 1909–10: "You see, I long for love, and I would so rejoice if I could hear an affirmative reply that I almost believe it is the fault of the damned science, so I gladly accept your smile on that account" (M-KS, 102).

It is unfortunate that, early in their marriage, Einstein did not recognize—or, if so, did not address —his wife's frustrations, bitterness, and sorrow following her failed exams, her lost career, and her lost child. Of course, he had his own setbacks, rejections, and burdens to bear, but he did have an advancing career and the support of his male friends. The University of Zurich post was only the first rung of the ladder Einstein climbed to reach the top of the physics profession. Propelled by the recognition he received and the need to support his growing family, he quickly moved from one university post to the next, each time uprooting his wife and boys. But there was also one frequent hurdle to overcome: as at the University of Zurich, it was the obstacle of anti-Semitism.

Already in March 1910 Einstein received an offer on the next rung of his career ladder: a "call" to an ordinary (full) professorship at the German Charles-Ferdinand University in Prague, which was then within Austria-Hungary. (A professorship was a calling, not a job.) The position also entailed appointment as director of the new institute for theoretical physics. "It is highly probable that I will be called to a large university as an ordinary professor with a significantly better salary than I now have," he wrote his mother in April 1910 (*CPAE* 5, doc. 204). On August 27, however, Einstein wrote Jakob Laub to say that Austrian bureaucrats imbued with anti-Semitic prejudice had refused to call him to the post (*CPAE* 5, doc. 224). But the bureaucrats finally relented, and Emperor Franz Joseph's culture minister issued the call to Einstein in January 1911 (CPAE 5, doc. 245). The Einstein family, including eight-month-old Eduard, made the move to Prague in March. Mileva was pessimistic about the move. "I cannot put it otherwise than to say that I am not going there gladly and that I expect very little pleasure from life there," she told Helene (M-KS, 104). Mileva's expectations were fulfilled. She recalled in a letter to Helene in December 1912 that although they had electricity for the first time and frequented a stimulating literary and music salon where they met Max Brod and, possibly, Franz Kafka, they were unhappy with the snobbery, poverty, unclean water, and lack of play areas for the children (M-KS, 107; Isaacson 2007, 166).

In addition, Mileva continually received reminders of her outsider status in science. For instance, when Einstein's friend the physicist Paul Ehrenfest came to visit, the Einsteins met him at the train station and accompanied him to a café. Not until Mileva left to care for the children did the café conversation turn to science (Isaacson 2007, 167). In October 1911, Einstein left his wife at home when he traveled to Karlsruhe, Germany, for another meeting of the Society of German Scientists and Physicians. "That must have been certainly very interesting in Karlsruhe," she wrote him longingly from Prague on

October 4. "I would have so gladly also attended a little and seen all those fine people." She signed it sadly "your old Dollie" (*CPAE* 5, doc. 290).

By late 1911, the Zurich Polytechnic, now conferring doctorates and newly renamed the Swiss Federal Institute of Technology, or ETH, had its eye on its alumnus Albert Einstein. But the bureaucrats needed convincing that a regular professorship in theoretical physics would not be a mere luxury. Einstein's friend Heinrich Zangger, a Zurich medical researcher, took up the cause, solicited letters of recommendation, and overcame any anti-Semitic objections. Mileva and the boys were overjoyed to move back to their beloved Zurich, especially to the ETH, for the start of the winter semester in October 1912. "We are all fine and we all look forward to Zurich," Einstein wrote his friend and collaborator Ludwig Hopf on June 12. "The ferocious Weber has died there, so it will also be personally very pleasant" (*CPAE* 5, doc. 408).

Weber had died in March. As only fair, Einstein was his successor and began offering during the winter semester the ETH's first lectures in the newly recognized discipline of theoretical physics. And he was now administering the diploma examinations on that topic. He was also hard at work on what would become, after several years of further very intensive work, his greatest achievement: the general theory of relativity. While the special theory encompassed the relative motions of observers at constant velocity, the general theory included accelerated motion. Einstein discovered that, because the effect of gravitation downward and acceleration upward on an object are physically equivalent, motion under a gravitational force in the vicinity of a massive object, such as earth, could be regarded instead as a curvature of space and time in the same region. But expressing this mathematically proved extremely difficult.

Einstein struggled to develop the theory with the mathematical help of his old friend and classmate Marcel Grossmann, who was now professor of mathematics at the ETH (see Pais 1982, chap. 12; Renn et al. 2007; Gutfreund and Renn 2015; Janssen and Renn 2015).

ENTER ELSA

In personal matters little had changed after the move. Despite her joy at returning to Zurich, Mileva repeated her familiar complaint in a December 1912 letter to Helene. It was the last surviving letter she wrote to her friend before a four-year gap in their correspondence that marked another difficult period for Mileva.

> We are well and are all, big and small, very happy to have turned our backs on Prague. … My big Albert has become a famous physicist who is highly esteemed by the professionals enthused about him. He is tirelessly working on his problems, one can say that he lives only for them. I must confess with a bit of shame that we are unimportant to him and take second place. We are living here happily and fairly quietly, since my husband does not like interruptions to his work. (M-KS, 107–108)

During the first of the famous Solvay congresses on physics, held in Brussels in late 1911, most of the world's leading atomic scientists gathered in Brussels to grapple with the appearances of the electron and the quantum. By virtue of his work on these subjects, Einstein was a leading participant. The "professionals" in attendance were so impressed that they invited him in April 1912 to visit Berlin for the first time. It was the beginning of Einstein's step up the last rung of his ladder to the top. But it was also the beginning of the end for Mileva.

During that April stay in Berlin, Einstein visited his relatives and his mother who, after the death of his father, had eventually moved to Berlin to be with her sister and her husband. This brought Einstein into renewed contact with his first cousin Elsa Löwenthal, née Einstein, whom he had last seen as a teenager. She was now thirty-six years old; he was thirty-three. Elsa, born in Hechingen in the Swabian region of southwestern Germany, had married a textile merchant, Max Löwenthal, with whom she had two daughters and a son who did not survive. In 1902 Löwenthal moved alone to Berlin after his business was liquidated. When he and Elsa divorced in 1908, she and her two daughters moved from Hechingen to an apartment above her parents' home in Berlin (*CPAE* 5, 640–641). Max Löwenthal died in 1914.

The age difference between Elsa and Albert was about the same as between Albert and Mileva, but Elsa was without career ambitions and quite comfortable in a domestic and nurturing role. Albert was smitten, as he wrote her from Prague on April 30, 1912, on the way back to Mileva in Zurich, of his inability to love his mother; then he declared, "But I have to have someone to love, otherwise life is miserable. And this someone is you; you cannot do anything about it, since I'm not asking you for permission. I am the absolute ruler in the netherworld of my imagination, or at least that is what I choose to think" (*CPAE* 5, Engl., doc. 389). But on May 21 he broke it off. "I am writing to you for the last time and put myself back into the unavoidable, and you must too. ... You know that I, like you, must carry my cross without hope" (*CPAE* 5, doc. 399). Einstein did not hear from Elsa again until nearly a year later. During that year the already emotionally fragile Mileva suffered a decline both physically and psychically. Surely realizing that living in Zurich would not solve her marital problems, Mileva sank deeper into depression, accompanied by the onset of rheumatism. She found it difficult to go out that winter on the icy streets of Zurich, and she began attending fewer and fewer of their once frequent musical evenings in the home of the ETH mathematics professor Adolf Hurwitz.

Einstein had given Elsa his office address. She wrote to him in March 1913 to wish him a happy thirty-fourth birthday and to ask him for a book about his work for general readers. Einstein responded on March 14 with an invitation to meet "whenever your way leads to Zurich." "Then we can take a nice walk (without my wife who is unfortunately very jealous), and I will tell you all about the remarkable things I have found" (*CPAE* 5, doc. 432). On May 23 he wrote, "What I wouldn't give if I could spend a few days with you, but without … my cross!" (*CPAE* 5, doc. 434, his ellipses). In October 1913 he did spend a few days with her in Berlin. Returning to Zurich he wrote on October 10 to tell her: "I now have someone of whom I can think with unbridled pleasure and for whom I can live." As for his wife: "She has not asked about you, but I believe she does not underestimate the significance that you have for me" (*CPAE* 5, doc. 476). Perhaps to keep the peace, a month earlier Einstein had traveled with his "cross" and their children back to Novi Sad to visit her family for a second time (*CPAE* 5, doc. 474).

THE CRISIS

Einstein's October 1913 visit to Berlin occurred in response to a visit by the two main chiefs of German physics, Max Planck and Walther Nernst, along with their wives, who had come to Zurich in July to make Einstein a fantastic offer. If all went well with the Prussian bureaucracy, they informed him, they could offer him an unparalleled dual post in Berlin as, first, a paid member of the prestigious Prussian Academy of Sciences; second, a full professorship at the Humboldt University of Berlin without the obligation of teaching. The latter position would later be replaced by the directorship of a new Kaiser Wilhelm Institute for Physics to be built as part of the newly founded Kaiser Wilhelm Society, a network of government sponsored research institutes (the predecessor of today's Max Planck Society). Although Einstein was indeed named director of the institute, its building was not constructed until 1936 with funds provided by the American Rockefeller Foundation, despite the Nazi regime then in power. By then Einstein had long since left Germany. (The institute later served as a center of German nuclear fission research during World War II.)

Einstein did not officially accept the Berlin offer until December 7, 1913 (*CPAE* 5, p. 635). During Christmas vacation later that month Mileva, now recovered somewhat, traveled to Berlin where she stayed with Albert's new colleagues, the chemists Clara Immerwahr Haber and her husband Fritz Haber. They helped her select a rental apartment for the family at Ehrenburgstrasse 33 in Berlin-Dahlem, not far from the Habers (Hoffmann 2013). But she knew the move to Berlin would not go well for her. The thought of Albert in the clutches of his mother and Elsa, along with his aunt and uncle nearby, must have weighed heavily upon her.

On March 21, 1914, Einstein left Zurich for good and headed alone to Berlin to take up his new position in April as the world headed toward war. He traveled first to Holland to visit relatives and Paul Ehrenfest. On March 29 he moved into the rental apartment and into an office at the nearby Kaiser Wilhelm Institute for Physical Chemistry, headed by Fritz Haber in Berlin-Dahlem. Einstein had left Mileva and the boys behind in Switzerland so that Eduard, who had been suffering from a long-term illness, could recuperate for several weeks at a Swiss spa.

Soon after Mileva and the boys arrived in Berlin around April 19, 1914, war erupted in the Einstein home, bringing the Bessos to Berlin from Italy. The breakup of a marriage is never pretty, nor are the ugly words and actions inflamed by over-heated emotions. According to a memo by Besso's wife Anna, written in 1918 after a conversation with Mileva, before her arrival Albert had sublet a room in the apartment without telling his wife in a crude attempt to force her out. During the ensuing uproar, which threatened a relapse of Mileva's illnesses, Clara Immerwahr Haber appeared and brought Mileva and the children to her home, while her husband took on the difficult role of mediator (*CPAE* 8B, appendix, 1032–1033). On Haber's advice Albert sent Mileva (still at the Habers with the boys) a sadly abusive and humiliating list of conditions under which he would allow her to live with him in the apartment. That he submitted them at all, he wrote, owed only to his desire not to lose the children (*CPAE* 8A, docs. 22, 23). The heartless list of demands, suited more for an indentured servant than a wife, revealed how far their marriage had deteriorated.

Conditions

A. You will make sure:

   1. that my clothes and laundry are kept in good order;
   2. that I will receive my three meals regularly *in my room*;
   3. that my bedroom and study are kept neat, and especially that my desk is left for *my use only*.

B. You will renounce all personal relations with me insofar as they are not completely necessary for social reasons. Specifically you will forego:

   1. my sitting at home with you;

2. my going out or traveling with you.

C. You will obey the following points in your relations with me:

1. you will not expect any intimacy from me, nor will you reproach me in any way;

2. you will stop talking to me if I request it;

3. you will leave my bedroom or study immediately without protest if I request it.

D. You will undertake not to belittle me in front of the children, either through words or behavior. (*CPAE* 8A, doc. 22; English from Isaacson 2007, 185–186, Einstein's emphases)

Remarkably, Marić agreed to such conditions and returned to the apartment to live with Einstein, probably for the boys' sake, while Fritz Haber negotiated a separation agreement (*CPAE* 8A, doc. 23). According to the initial agreement, Marić would return to Zurich with the boys; Einstein would see the boys only outside his home and without Elsa; and he would support her and the boys financially. In order to lend legal status to the agreement, Haber arranged for an official signing of the document at a lawyer's office. Marić and Haber appeared, but Besso represented Einstein (*CPAE* 8B, appendix, 1032–1033; Isaacson 2007, 186–187).

Einstein later attempted to explain to Helene Kaufler Savić, using Mileva's nickname: "Separation from Mitsa was for me a question of life. Our life in common had become impossible, even depressing, but I could not say why. So I am giving up my boys, whom I love so tenderly" (M-KS, 109–110).

On July 28, 1914, Austria-Hungary declared war on Mileva's homeland Serbia, igniting World War I. On the following day Mileva and the boys, accompanied by Besso, boarded a train back to Zurich. Einstein, appearing at the Berlin train station

Figure 3.1
Eduard, Mileva, and Hans Albert in Berlin, July 1914. Photo: Albert
Einstein Archives © The Hebrew University of Jerusalem, courtesy of
the Emilio Segrè Visual Archives, American Institute of Physics.
Permission: Courtesy of the Leo Baeck Institute, New York.

with Haber, was devastated by the loss of his children. He
reportedly cried all afternoon (Isaacson 2007, 187). That evening he stayed at Elsa's parents' home (Elsa was conveniently
absent), and the next day at his mother's home. The marriage
was over. Walter Isaacson writes:

> The decline of the marriage was a downward spiral. He had become
> emotionally withdrawn, Marić had become more depressed and
> dark, and each action reinforced the other. Einstein tended to avoid
> painful personal emotions by immersing himself in his work. Marić,
> for her part, was bitter about the collapse of her own dreams and
> increasingly resentful of her husband's success. Her jealousy made
> her hostile toward anyone else who was close to Einstein, including
> his mother (the feeling was reciprocal) and his friends. Her mis-

trustful nature was, understandably, to some extent an effect of Einstein's detachment, but it was also a cause. (Isaacson 2007, 184–185)

TRIUMPH AND TRAGEDY

Einstein did not at first seek a divorce. He maintained his own apartment, visited Elsa often, and, above all, continued his devotion to his work. He rebuffed Elsa's many entreaties regarding divorce and remarriage. In a letter to Mileva on December 12, 1914, he specified his financial support of her and the children in the amount of 5,600 German marks per year in quarterly payments, which was about 47 percent of his Berlin salary of 12,000 marks, paid through the state-sponsored Prussian Academy of Sciences (*CPAE* 5, doc. 485; 8A, doc. 40). To put this in perspective, according to data compiled by the German Central Bank (Deutsche Bundesbank) and consulted by Fölsing (1997, 421), the prewar German mark (M) to Swiss franc (SF) exchange rate was 83 M to 100 SF, making 5,600 M equivalent to 6,747 SF. This amounted to about 61 percent of his last annual salary of 11,000 SF as professor of theoretical physics at the ETH (*CPAE* 11, pp. 189, 192).

In November 1915, Einstein submitted to the Prussian Academy for publication in its *Meeting Reports* (*Sitzungsberichte*) the first of his papers on one of the greatest achievements of the century, the theory of general relativity—a theory that encompassed gravitation, with wide-ranging applications to cosmology and the origin of the universe, the prediction of gravitational waves, and such earthly inventions as today's global positioning system (*CPAE* 6, docs. 21–22, 24–25). At the same time, Elsa and her parents kept up the pressure on Einstein to divorce Mileva and marry Elsa by raising the argument that Elsa's two daughters, who were approaching adulthood, would find it difficult to marry within proper Berlin society if their mother, a divorced woman, even with a deceased ex-husband, was consorting with a married man; nor would it look good for the famous professor.

Einstein finally relented. In February 1916 he wrote Mileva proposing that they turn their separation into a divorce, with their Berlin separation agreement as the basis for the divorce agreement. He also offered her, in addition to the 5,600 M, an additional 6,000 M for child support left over from his Prague salary (*CPAE* 8A, docs. 187, 200). Mileva refused. By July she was in a state of collapse, bedridden by heart ailments, rheumatism, and extreme anxiety. Heinrich Zangger, the Zurich medical researcher and close friend of the Einsteins, moved her to a sanatorium and arranged for a governess to care for the boys. The Bessos soon stepped in to replace the governess, followed by Mileva's mother. Finally, Mileva's old friend Helene Kaufler Savić, who had moved near Lausanne with her two girls for the duration of the war, took charge of the children in Lausanne. Her husband was away procuring supplies for the Serbian army (Isaacson 2007, 230–232; *CPAE* 1, 386).

Mileva's health improved a year later, when Einstein withdrew his divorce request. Eventually she was able to care for the boys again. But then Einstein began to suffer gastrointestinal problems, apparently related to an ulcer (*CPAE* 8B, ed. note, 817). In order to provide him better care, Elsa arranged for him to rent an apartment in the building where she and her family lived.

On January 31, 1918, Einstein formally requested a divorce for the second time (*CPAE* 8B, doc. 449). By then the continuing war had driven down the value of the mark relative to the Swiss franc. Switzerland had remained neutral. At the exchange rate of 135 M to 100 SF, Einstein had to pay 9,108 M annually in order to maintain the same amount of support as before (6,747 SF) (Fölsing 1997, 421). With the mark likely to drop further and his family's mounting expenses from hospital and spa care. Einstein, Marić, or a financial adviser, came up with the idea that the financial award from the Nobel Prize, should he receive it, would help greatly in meeting his future support obligations. Utilizing legal language in numbered paragraphs that suggest consultation with a lawyer, Einstein wrote to Mileva on January 31, 1918, that, first, he would pay her 9,000 M per year, of which 2,000 M would be deposited as child support. Second, he wrote, "The Nobel Prize—in case of the divorce and in the event it is awarded to me—would be given over [*abgetreten*] completely to you a priori." That he was referring to the Nobel Prize money rather than to the prize itself is indicated by the next sentences, which also clarify what he meant by "completely": "The free use of the interest would be given to you [*bliebe Dir überlassen*]. The principal would be deposited in Switzerland and held in security for the children." And, he added, in this scenario, rather than the fixed 9,000 M per year, the interest and his payments together would always amount to a minimum of 8,000 M per year (*CPAE* 8B, doc. 449).

Although Mileva was recovering from her illnesses when she received Einstein's letter, she was again burdened with caring alone for Eduard, who now suffered from inflamed lungs and required recuperation at a sanatorium in Arosa. In addition, Mileva's sister Zorka, who had come to Zurich to care for her sister, promptly suffered a mental breakdown. She too had to be hospitalized. Then their brother Miloš, who was a medic in the Serbian army, was missing in action on the Russian front. It turned out he was captured by the Russians and taken east as a prisoner of war. He survived the Russian Revolution but never returned. In the midst of all this, the prospect of peace of mind and sufficient financial support apparently convinced Mileva to accept Einstein's request. She commissioned her neighbor, a lawyer, to represent her (*CPAE* 8B, doc. 457). Again, she insisted on keeping the children in Zurich. With the loss of her career and her child and now her marriage, her two boys were all she had left. On June 7, 1918, Mileva signed the divorce agreement in Zurich. Einstein signed it five days later in Berlin, thereby reluctantly acceding to her wishes regarding the children (*CPAE* 8B, doc. 562). But the divorce had to be finalized by a Swiss court in Zurich, since they had married in that country nearly sixteen years earlier.

DIVORCE

On February 14, 1919, the Swiss court finally issued the divorce decree and confirmed, along with it, their final agreement regarding support and the Nobel Prize money (*CPAE* 9, doc. 6). It required, however, Einstein's open admission of one of the few permitted grounds for divorce at that time: adultery. He married Elsa three and a half months later.

The financial provisions in the divorce agreement and final decree were close to Einstein's original offer, but they were now worked out in greater detail. According to documents published in the *Collected Papers* in 1998 and 2004 (*CPAE* 8B and 9), upon signing the 1918 agreement Einstein had deposited 40,000 M in a Swiss bank account under Mileva's name from which she received the interest toward alimony and child support, but she could not withdraw from the principal without Einstein's approval (*CPAE* 8B, doc. 562). Then, should he subsequently receive the Nobel Prize, he would deposit the prize money in a Swiss bank, minus the 40,000 M he had already deposited. The total trust would thus amount to the full prize money in her name as her "property" (*Eigentum*). But again she, and the boys through her, would have free access only to the interest; his permission was required to access the principal. In both cases he would ensure, through additional payments if necessary, a minimum income of 8,000 SF per year (not German M as earlier, the value of which was still dropping). Upon her death or remarriage, the trust would become the children's property. In the meantime, Mileva received full custody of the children. They could visit with their father only when he came to Switzerland and only during school vacations (*CPAE* 8B, doc. 562; *CPAE* 9, doc. 6).

Einstein finally did receive the Nobel Prize for Physics in November 1922 (for 1921). The financial award amounted to about 120,000 Swedish kronor, or (avoiding the German mark then teetering on collapse) about 180,000 SF (Fölsing 1997, 542). Upon Mileva's request, he give her 45,000 SF for the down payment on a multifamily apartment building in Zurich where she and the boys would reside and utilize the rents received as income and payment of the mortgage for the remainder of the cost (*CPAE* 14, doc. 37). Einstein deposited another 45,000 SF in a Swiss bank under her name. But over her vehement objections, and in violation of the divorce agreement, he placed the remainder of the prize money (90,000 SF) under her name with a New York investment firm, rather than with a Swiss bank. The postwar American economy was booming—and it seemed safe until the 1929 crash. Despite the divorce agreement violation, Einstein told his former wife that he had now fulfilled his obligations (*CPAE* 14, docs. 204, 205; Milentijević 2015, chap. 18; Wolff 2016).

Einstein married his cousin Elsa in Berlin on June 2, 1919. One month earlier an observation team led by the British astronomer Arthur Eddington famously detected the bending of starlight by the curved space near the sun, exactly as predicted by Einstein in his general theory of relativity. Upon Eddington's report to a joint meeting of the Royal Society and Royal Astronomical Society later that year, Einstein suddenly became a world celebrity.

Mileva spent the rest of her days in Zurich, often ill and caring for her son Eduard, who suffered from schizophrenia. Hans Albert became a noted civil engineer and a source of great pride to his mother. He later became a professor at the University of California at Berkeley, and his son Bernard Caesar became a physicist. Mileva's long-time friend and confidant Helene Kaufler Savić managed to remain in Belgrade under the German occupation during World War II, even though she was half Jewish. She died of starvation and pneumonia in 1944 (M-KS, 26). Her daughter Julka became a medical doctor and professor. Mileva Einstein-Marić never remarried. She died in Zurich on August 4, 1948, at age seventy-two.

As we have seen in this brief section, a properly researched history reveals the real Mileva Einstein-Marić to be a woman of great promise, determination, and courage as an international pioneer for women in science at the turn of the century. But in the end, she did not experience the success she had hoped for in physics or in mathematics or in life. As Ellen Goodman wrote:

> The tragedy of Mileva's life was real enough. But it's of a more personal and common dimension. It's a parable of two young people who begin life as intellectual soul mates. ... But somewhere along the way, life and love had an unequal effect in their lives as man and woman, and as scientists. (Goodman 1990)

II WOMEN IN SCIENCE

RUTH LEWIN SIME

.

Mileva Marić's story resonates with us, and for good reason. We recognize the determined girl in a school for boys, the confident student who set her sights on a career in physics, the passionate young woman who wanted an egalitarian marriage. We understand that Mileva was pushing the limits of what girls and women were allowed to do, and that she and women like her succeeded in bringing about change for others. By the turn of the twentieth century, Germany and Austria-Hungary, the Western European states most resistant to the idea of higher education for women, finally opened their universities to women students. Education was an essential step toward equality for women, but it was only the first. If a woman wanted to pursue a career in science, there was no obvious path for her to follow. Even the most brilliant and ambitious women encountered formidable social and institutional barriers that would remain in place for generations, with effects that are still noticeable today.

It is important to remember that there have always been women who were scientists and mathematicians. Long before women had access to universities or even much schooling, there were individuals of great creativity, even genius, who were acclaimed by their peers and remembered by historians; the list includes Laura Bassi, Émilie du Châtelet, Sophie Germain, Mary Somerville, Sofia Kovalevskaia.[1] Most women scientists learned from family members and worked in domestic surroundings, because that is where most scientific practice took place.[2] Caroline Herschel, for example, assisted her brother with his astronomical observations and became an important astronomer in her own right.[3] Maria Mitchell learned astronomy from her father and gained international recognition for discovering a new comet, the first American to do so.[4] Marie Anne Paulze Lavoisier worked with her husband in the chemistry laboratory in their home, their partnership captured in a stunning double portrait by Jacques-Louis David.[5] Historians have uncovered many such women; very likely there are many more whose names we do not yet know.

In the nineteenth century scientific practice became increasingly professionalized, and more gender-segregated. To the extent that a scientist was expected to be a man with a full-time paid position outside the home, science itself came to be seen as intrinsically masculine, and women, however well educated, were left out.[6] Women occupied whatever scientific niches were available to them, often in jobs that were essential but regarded (and paid) as if they were peripheral. They worked as illustrators, librarians, textbook writers, technicians, and assistants.[7] They were professors in women's colleges, where they taught, did research, and campaigned for women's rights.[8] They were hired by observatories, industrial laboratories, and government facilities as human "computers," in groups that were entirely segregated by gender and at times by race.[9] They were recruited for the Manhattan Project and then dismissed when the war was over.[10] They clustered into research fields that were more receptive to women, including radioactivity, X-ray crystallography, and astronomy.[11] And, quite often, they joined their husbands in collaborative scientific marriages, much as Mileva and Albert envisioned in their student days.[12]

One of the most successful and visible marital collaborations was that of Marie and Pierre Curie. In 1903, when they were awarded the Nobel Prize in Physics (with Henri Becquerel) for their discoveries in radioactivity, the public reaction was a mixture of excitement and disbelief that a woman, wife, and mother could be an important scientist.[13] By the time the Czech American biochemists Gerty and Carl Cori were awarded the 1947 Nobel Prize in Physiology or Medicine (with Bernardo O. Houssay) for their studies of carbohydrate metabolism, high-level women scientists were no longer a complete rarity, but it is notable that for both Nobel couples, in France and the United States, some forty years apart, it was the husband who had position and status while the wife made do in marginal jobs.[14] This pattern persisted for much of the twentieth century, and it applied to couples who did not collaborate directly, such as the German American theoretical physicist Maria Goeppert-Mayer and her American husband Joseph Mayer, a physical chemist. For the first thirty years of their marriage Joe was a distinguished university professor while Maria held a series of adjunct positions, even though she was internationally recognized for her theoretical work in quantum physics and nuclear physics. In 1960 both Joe and Maria accepted professorships at the University of California, San Diego, and in 1963 Maria was awarded the Nobel Prize in Physics (with J. Hans D. Jensen) for the discovery and development of the nuclear shell model. A local newspaper headline, "La Jolla Mother Wins Nobel Prize," shows just how unimaginable it still was, sixty years after Marie Curie's first Nobel Prize, for the words "scientist" or "physicist" to be used when referring to a woman.[15]

We know less about unsuccessful scientific couples, but one exceptionally tragic instance is quite well documented. When

Clara Immerwahr married the physical chemist Fritz Haber she had a doctorate in physical chemistry from Breslau University (in 1900, eight years before women were officially accepted as students), several publications, and an intense desire to continue scientific research. Clara hoped she and Fritz would be able to work together, but that did not take place and their marriage was unhappy. In 1915, when Fritz was home in Berlin for a few days on leave from the war, Clara killed herself with his service pistol. Although some writers have interpreted Clara's suicide as a protest against her husband's leadership in developing poison gas warfare, existing documents indicate that she had long been despondent about her failed marriage and her loss of identity as a scientist.[16]

Of the first generation of university-educated women scientists, many chose to remain single in order to devote themselves to their work. Even so, relatively few had untroubled careers, even when their achievements were extraordinary. Emmy Noether is regarded as one of the most important mathematicians of the twentieth century, known for her far-reaching innovations in modern abstract algebra and for profoundly influencing the development of modern physics with her solution to the problem of energy conservation in Einstein's general theory of relativity. For most of her twenty-five years of professional life in Germany, primarily in Göttingen, she did not have a paid university position, despite the efforts of her colleagues. For years the university would not allow her to undergo *Habilitation,* the qualification process for an academic position, for no reason other than her gender. It was not until 1933, when she was forced to emigrate due to Nazi racial policies, that Noether had a position with a decent salary. She accepted a professorship at Bryn Mawr, a women's college near Philadelphia, and maintained contact with Einstein and other colleagues in Princeton until her untimely death in 1935.[17]

The Austrian physicist Marietta Blau pioneered the use of photographic emulsions for imaging high-energy nuclear particles and events. In 1937 she and a co-worker, Hertha Wambacher, found that heavy nuclei in the emulsion had been explosively disintegrated by cosmic radiation, a discovery that dazzled the nuclear physics community and launched the field of particle physics. Blau worked in Vienna at the Institute for Radium Research, where more than a third of the scientists were women. Although none of the researchers had regular paid positions, the men could undergo "habilitation" while very few women did; when Blau inquired about it she was told, "Woman *and* Jew: that is too much." Her lack of academic standing made her emigration difficult. After leaving Vienna in 1938 she was unable to do research in her field for the next ten years, and by the time of her death she and her work had been almost entirely forgotten.[18]

Lise Meitner, a native of Vienna, made her professional home in Berlin, where she worked in radioactivity, pioneered the new field of nuclear physics, and, with the chemists Otto Hahn and Fritz Strassmann, was a co-discoverer of nuclear fission. Meitner's interdisciplinary partnership with Hahn was essential for her start, giving her a place to work and a chance to prove herself at a time when women had no prospects for a position in academic research. Once underway, her career was extraordinary: in 1913, a permanent appointment in the Kaiser Wilhelm Institute for Chemistry; in 1917, her own section for physics in the Institute, the equivalent of a university professorship; habilitation in 1922 and an adjunct professorship at the University of Berlin in 1926—each step a milestone for the inclusion of women into German science. In the 1920s Einstein liked to call her "our Marie Curie," an acknowledgment of Meitner's importance in the Berlin physics community and her international prominence. When she escaped from Germany in 1938 she went to Stockholm, where she was given an entry-level position without a laboratory or the resources to work as she had before. Exile shattered Meitner's career, diminished her reputation, kept her from a Nobel Prize, and eventually clouded her place in the history of twentieth century science.[19]

Of the early generations of women scientists, nearly all experienced difficulties in their professional lives and many were severely under-recognized. Women scientists were doubly outsiders: as women they were unconventional, as scientists they were not fully accepted in a traditionally male domain. That left them vulnerable, not among the close friends and colleagues who knew them and understood their work but in the wider scientific community and beyond. This was Meitner's situation in Sweden, where she was viewed primarily as a woman and a foreigner rather than as a prominent physicist with multiple Nobel nominations who could contribute to Swedish science. Marie Curie was similarly vulnerable. After the 1903 Nobel Prize, some imagined that she was just her husband's helper or, more romantically, his muse. When Pierre Curie died in 1906 and Marie was appointed to his professorship in the Sorbonne, there were people who assumed that widowhood was her main qualification for the position. And in 1911, just as she was awarded her second Nobel Prize—this time in chemistry, for the discovery of polonium and radium, and the isolation of elemental radium—she was being vilified in the press because of her affair with a married colleague. Curie was never elected to the French Academy of Sciences; it took years before she became the iconic figure everyone recognizes today.[20]

It took generations before women scientists began to be perceived as normal, both as women and as scientists. The process is still underway. As recently as 2005 the president of Harvard University publicly speculated that "issues of intrinsic aptitude" might be holding women back.[21] We are still worrying about the paucity of women and minorities in mathematics, the physical sciences, engineering, and technology, still wondering when everyone with the ability and desire for a life in science will have the opportunities they need to succeed. Looking back, we seek a deeper understanding of the many women who also dared to hope and work for a life in science, as did young Mileva.

III EXAMINING THE MILEVA STORY

ALLEN ESTERSON

5   THE STORY BEGINS

The Mileva Story has its roots in an oral tradition extending back to relatives and friends of the Marić family. Some of these stories were later collected and published around ninety years later (Marković 1995). The earliest known published claim regarding Marić's role as a contributor to Einstein's work appeared in an interview published on May 23, 1929, with Marić's close friend from her Zurich days Milana Stefanović (née Bota), a former Zurich psychology student from Serbia.

THE 1929 INTERVIEW

The interview to be examined below has been cited to support the contention that Marić collaborated with Einstein on his 1905 relativity paper. It was conducted by the journalist Miša Sretenović and appeared in a Belgrade publication, but authors differ on which publication.[1] Wherever it appeared, the interview and its subsequent uses exemplify the difficulties in analyzing and interpreting such a source: the mix of languages, the translations tailored to preconceptions, and the unreliability of hearsay and memory. None of these, however, hindered subsequent authors from using this interview as an important source for the Mileva Story.

Outside of Serbia, the only available versions of the key passage to support the later Mileva Story come to us as a three-sentence excerpt from the original Serbian, followed by German and English translations offered by three authors. Desanka Trbuhović-Gjurić, who extracted the three sentences in her biography of Marić (T-G 1969), provided the Serbian. A German translation of the sentences, probably by Trbuhović-Gjurić herself, later appeared in the German editions of her book (T-G 1983, 75–76; 1988, 93–94). Trbuhović-Gjurić's Serbian excerpt was also translated into English by Karlo Baranj and published by Michele Zackheim in her book *Einstein's Daughter: The Search for Lieserl* (Zackheim 2000, 20, 282). The Serbian author Dord Krstić provided a second English translation in his book *Mileva & Albert Einstein: Their Love and Scientific Collaboration*, in which he cited the original interview article as his source (Krstić 2004, 122). Krstić's English translation (included below) agrees more closely with the German translation.

As reported by Sretenović, Milana Stefanović (to avoid confusion, we refer to her henceforth as Milana) was evidently responding to a question about Einstein's work when she stated:

> Mica [variant of Mileva's nickname] would be the most relevant one to give information about the genesis of his [Einstein's relativity] theory, since she was also involved in its creation. Five, six years ago, Mica told me about that, but with pain. Perhaps it was difficult for her to evoke the memories of her most pleasant hours, or maybe she didn't want to "reduce" the great glory of her ex-husband. (Krstić 2004, 122)

In Zackheim's English translation, the last clause reads: "maybe she does not wish to kill the great voice of her former husband" (Zackheim 2000, 19, 282). Trbuhović-Gjurić's German translation is more prosaic: "vielleicht wollte sie auch dem grossen Ansehen ihres einstigen Mannes nichts anhaben" (T-G 1983, 75–76; 1988, 93–94); that is, "perhaps she did not wish to harm the great reputation of her former husband" (my translation).

Sretenović's report, along with quotations from an unscholarly source (see the section on Peter Michelmore below), served in that chapter in Trbuhović-Gjurić's influential book as evidence that Marić was involved in the formulation of Einstein's special relativity theory. But, as typical of hearsay evidence, we cannot know what Marić actually told Milana. Moreover, Milana's vagueness (as reported by the journalist) does not indicate a definite statement on Marić's part. The words "[m]aybe it is hard for her to recollect those most precious moments, maybe she does not wish to kill the great voice of her former husband" (Zackheim 2000, 20) suggest that Milana was inferring more than Marić actually told her.

Marić herself did not support Milana's claims in the interview, if indeed she was fully aware of them. According to Krstić (2004, 122), Marić had gone to her hometown of Novi Sad in southern Hungary in 1929 to care for her ailing mother. She then visited another close friend from her Zurich days, Helene Kaufler Savić, who resided near Belgrade, Serbia. The visit is confirmed by a letter Marić wrote to Kaufler Savić on June 13, 1929, after returning to Zurich (M-KS, 157). This was about three weeks after the publication of the interview in Belgrade. It is not certain if Marić saw the interview or if Milana simply told her about it via letter, but she referred to it in her June 13 letter to Kaufler Savić:

> Milana wrote me a very self-assured letter. ... Milana could not help confiding our stories to the newspaper reporter, and I thought then that the matter was finished, so I did not talk about it at all. I would avoid being involved with such newspaper publications, but I believe it gave pleasure to Milana, and she probably thought that it would give me pleasure, too, and in a way would help me to acquire certain rights vis-à-vis E. [Einstein] in people's eyes. (M-KS, 158)[2]

It is evident from this letter to Savić that Marić neither denied nor supported the version of the story that Milana passed on to the journalist—a version Milana claimed Marić told her five or six years earlier regarding her role in the formulation of the special theory of relativity. If, as seems likely, Marić had not seen the actual article in the Belgrade newspaper, but relied instead on what Milana wrote in her letter, it is possible Marić was not fully aware of what her friend had told the journalist.

Even though the interview content is wholly unreliable as hearsay, it is still of interest to consider the objectivity of the source (or her lack of it). In this case, by the end of her student days in Zurich, Milana Bota (then single) had come to dislike Einstein. As Highfield and Carter (1993, 110) observe: "Milana Bota was not a disinterested party; her account reflects her great affection for Mileva, and perhaps also her lingering resentment against Einstein." Einstein was a frequent visitor to Pension Engelbrecht in Zurich where Marić was lodging along with Milana Bota, Helene Kaufler (then single), and other young women who had come to Zurich for higher education. He often joined them with his violin for musical sessions, and Bota wrote highly of him to her parents in 1898 (M-KS, 4). However, as the romance grew, and as Einstein came to monopolize Marić's time (they also regularly read books by eminent physicists together), Bota's attitude changed radically. In July 1900 she wrote to her mother: "I see little of Mitza [variant nickname] because of the German [Einstein], whom I hate ..." (*CPAE* 1, doc. 64, n.6, her ellipses; T-G 1983, 55; 1988, 64, with ellipses removed). No doubt Einstein's propensity to tease Marić's friends added extra force to Milana's resentment against him, whatever the reason or the nature and depth of her resentment (*CPAE* 1, doc. 64, n. 6; M-KS 57).

Twenty-nine years later, Milana may have still harbored the
feelings expressed in her student days. Whether or not this was
the case, as a friend of Marić's not only was she not a disinterested interviewee, but also the isolated three-sentence interview
excerpt is plagued by all of the shortcomings of hearsay evidence. Despite these caveats, as a result of the excerpt in question being included in Trbuhović-Gjurić's book, it was regarded
by those willing to overlook its obvious difficulties as the earliest
piece of written evidence in support of the Mileva Story.

HANS ALBERT EINSTEIN AND THE MICHELMORE EPISODE

In 1962, the distinguished Australian journalist Peter Michelmore, then working as a foreign correspondent in the United
States, published his short biography *Einstein: Profile of the Man*
(Michelmore 1962, British edition 1963; the page numberings
differ). In preparation for the work, he traveled to California to
visit and interview Einstein and Marić's son Hans Albert Einstein for two days in February 1962. Hans Albert was born in
Bern, Switzerland, in May 1904 and was by 1962 an engineering
professor at the University of California, Berkeley. In the context
of Einstein's 1905 relativity paper, in his book Michelmore
stated, without citing a source: "Mileva helped him solve certain
mathematical problems, but nobody could assist with the creative work, the flow of fresh ideas" (Michelmore 1963, 41).

It might be assumed that the source of this statement was
Hans Albert and it is conceivable that Hans Albert said something of that nature. But in the absence of Michelmore's interview notes we have no idea what Hans Albert actually told him
fifty-seven years after the event. Since Michelmore did not attribute the statement to anyone, it could just as easily have been

the product of his imagination, as were many other items in his book (see below). In any case, as the science historian Alberto Martínez has observed, Hans Albert was only a baby in 1905, so anything he might have said about his mother helping with the mathematics in that year could not have come from his first-hand knowledge (Martínez 2009, 201).

Two of the subsequent founders of the Mileva Story, Trbuhović-Gjurić and Senta Troemel-Ploetz, embellished Michelmore's account by implying a more direct source: Einstein himself. According to Trbuhović-Gjurić:

> Peter Michelmore, who had much information from Albert Einstein, said: "Mileva helped him solve certain mathematical problems." (T-G 1983, 72; 1988, 90)

Troemel-Ploetz, citing Trbuhović-Gjurić's 1983 work, included her Michelmore passage verbatim, but took it further:

> Peter Michelmore, who had much information from Albert Einstein, said [ref in original: T-G 1983, 72]: "Mileva helped him solve certain mathematical problems. She was with him in Bern and helped him when he was having such a hard time with the theory of relativity." (T-P 1990, 419–420)

The two sentences Troemel-Ploetz quotes from Trbuhović-Gjurić's book appear contiguously in her article but in fact come from separate pages (T-G 1983, 72 and 74; 1988, 90 and 92). In both cases, the authors implied that this information came to Michelmore directly from Einstein, which, to unsuspecting readers, was not inconceivable since Einstein had died only seven years earlier. But more importantly, neither of them took the trouble to check the source and find that Michelmore (1963, ix) did not acknowledge Albert Einstein in his "Author's Notes."

MICHELMORE'S IMAGINATION

There is no way of knowing how much of what Michelmore wrote in his book came from Hans Albert, but—surprisingly for such a distinguished journalist—some of it clearly came from Michelmore's own imagination, as well as, on occasion, from inaccurate information provided by Hans Albert. That imagination is on full display in numerous fictional scenarios in his book with invented dialogue, presumably to make his narrative more novelistic (e.g., Michelmore 1963, 6, 10, 21–22, 28–29, 31–32, 33–34, 35, 36, 37, 39–40, 42–43, 46, 49). To take just one example:

> After a year in Zürich, Einstein was offered a full professorship in experimental physics at the German University in Prague. It was tempting. The salary was nearly double the twenty-four pounds a month he was being paid in Zürich.
>
> "But it's not your field," argued Mileva. "You're no good at experiments."
>
> "No, you're right. I'll turn it down."
>
> "It's a big break, Albert," a colleague told him. "You can't afford to pass up a full professorship."
>
> "Of course not. I'll take it."
>
> Einstein, as usual, was too involved with his private theoretical work to form an opinion on whether or not to uproot his family and move to Prague. But the decision was important to Mileva. She demanded that he make up his mind. Finally, he decided to give Prague a try. (Michelmore 1963, 49)

Einstein's reputation was based entirely on his work in theoretical physics. He was not offered a professorship in *experimental* physics but in theoretical physics, and he became the director of a new institute in that field (Frank 1947, 98).[3] That alone, aside from the obvious doubt that such a conversation between Mileva and Albert actually took place, undermines any notion that the Michelmore biography of Einstein is a serious work of scholarship. Nevertheless Trbuhović-Gjurić quotes much of the scenario from Michelmore (1963, 49), translated from the English with heavy editing, as if it were historical fact, and without any acknowledgment (T-G 1983, 89–90; 1988, 107–108). At least two other such instances from Michelmore (1963, 31–32, 42) appear in her book (T-G 1983, 66, 72; 1988, 84, 90).

MICHELMORE ON MARIĆ'S MATHEMATICS

After correctly noting that in his later years as a Zurich Polytechnic student Einstein felt he knew enough basic mathematics for his purposes, and thus sometimes skipped mathematics lectures, Michelmore states:

> It was a fortunate accident that Einstein's closest friends at the institute were both studying mathematics. One was Marcel Grossmann who was genuinely awed by the range of Einstein's mind. … Generously, Grossmann took detailed notes on all lectures and drummed them into Einstein at the week-ends. … His other close friend was Mileva Marić … She was as good at mathematics as Marcel, and she, too, helped in the week-end coaching sessions. (Michelmore 1963, 31)

Other than the fact that Einstein made use of Grossmann's meticulous lecture notes, this scenario is complete fiction. Einstein borrowed Grossmann's notes when he was preparing for the intermediate and final diploma examinations in 1898 and 1900, respectively (Einstein 1956a, 11; Pais 1982, 44; Fölsing 1997, 56, 68). The absurd story (and picture) of drumming his notes into Einstein's head, and of Marić assisting in the effort, is found only in Michelmore's book.

Likewise, the notion that Marić was "as good at mathematics as Marcel" is at variance with the grade reports now available. Grossmann achieved higher grades in all four mathematics subjects for which they were tested on the intermediate and final diploma examinations (*CPAE* 1, docs. 42, 67; and T-G 1988, 63, see appendix D). He became professor of geometry at the Polytechnic in 1907, and in 1911 he succeeded to the directorship of the mathematics-physics section, Section VI A (*CPAE* 1, 381). In contrast, Marić twice failed the Zurich Polytechnic final diploma exams largely as result of her poor grades in the mathematics component (see appendix D). She left no known mathematical achievement beyond her Polytechnic work.

Evidently Michelmore had no knowledge of Marić's rather mediocre mathematical performance at the Polytechnic, which may indicate that he was unaware of Einstein's performance as well. Yet in the popular literature his unsubstantiated assertion that Marić was as good at mathematics as Grossmann was so readily accepted as historical fact that a chain of citations developed, starting with Dord Krstić's citation of Michelmore in 1991:

> It is unlikely that her [Mileva's] contribution to Albert Einstein's work will ever be determined precisely. However, if we keep in mind that "she was as good at mathematics as Marcel (Grossmann)," we may suppose that her part was not small [ref in original: Michelmore 1962, 35]. (Krstić 1991, 98)

In 1995, Andrea Gabor wrote, citing Krstić (1991, 98) but misspelling Grossmann's name:

> [Around 1913] Einstein began collaborating with Marcel Grossman on the general theory of relativity; the collaboration is particu-

larly noteworthy since, according to Einstein's biographer Peter Michelmore, Marić was "as good at mathematics as Marcel [Grossman]." (Gabor 1995, 25)

More than a decade later Edith Borchardt (2008, 6) repeated the quotation directly from Gabor, with the same misspelling of Grossmann's name. Likewise, after stating that "Michelmore was the first of Einstein's biographers to acknowledge … that she played a significant role in Einstein's achievements of 1905," Radmila Milentijević (2015, 121) repeated the same quotation (citing Trbuhović-Gjurić) as if it constituted historical fact.

6   THE STORY EMERGES

The Mileva Story reached its full fruition in the Serbian and German editions of Trbuhović-Gjurić's biography of Mileva Einstein-Marić (T-G 1969–1995), the German translations being published by Paul Haupt. This hitherto little-known work came to wider public attention with the 1987 publication of the first volume of *The Collected Papers of Albert Einstein* (*CPAE*) and the newly discovered Einstein-Marić so-called "love letters" it contained. (Items in the letters became a secondary source of claims about the supposed contribution of Marić to Einstein's early scientific achievements.) Three years later the German linguist and psychotherapist Senta Troemel-Ploetz, who had not previously written on Einstein or other historical topics, brought Trbuhović-Gjurić's story to English-speaking readers in the United States and abroad through her lengthy article "Mileva Einstein-Marić: The Woman Who Did Einstein's Mathematics" (T-P 1990). In it Troemel-Ploetz drew heavily upon the 1983 German edition of Trbuhović-Gjurić's biography, as well her own conclusions drawn from the Einstein-Marić correspondence in *CPAE* 1.[1] For subsequent authors who were not able to read Trbuhović-Gjurić's Serbian or German editions, or the French translation of 1991 (it has not been published in English), Troemel-Ploetz offered a readily available resumé of the book.

DESANKA TRBUHOVIĆ-GJURIĆ

Desanka Trbuhović-Gjurić was born in 1897 into an educated Serbian family living in Krapina, Croatia. She died in Belgrade, Serbia, in 1982, the same year the first edition of the German translation of her book was published. According to Werner G. Zimmermann, a former Zurich city archivist and the editor of the German editions of her book (which Trbuhović-Gjurić herself probably translated into German), Desanka Gjurić (or Djurić) followed the career to which Marić had aspired (Zimmermann 1988b, 209–210). She studied mathematics and physics at the universities of Zagreb and Prague. Subsequently she became a gymnasium (high school) teacher in those subjects and married Borislav Trbuhović in 1922, placing his last name before hers, as did Mileva Einstein-Marić.

Those were the early years of the independent Kingdom of Serbs, Croats, and Slovenes following World War I. The kingdom became the Kingdom of Yugoslavia in 1929. After World War II, Trbuhović-Gjurić taught at the technical college and university in Belgrade. Upon retiring, she devoted herself to researching and writing about the woman of whom "there is every reason to be proud that she belonged to our [Serbian] people," and whom she revered as "great in many ways": Mileva Einstein-Marić (T-G 1983 and 1988, 7).

Trbuhović-Gjurić's research led in 1969 to the publication of her biography of Marić in Serbian, *U senci Alberta Ajnštajna* (In the Shadow of Albert Einstein). In 1982 the first German translation appeared with the title *Mileva Marić-Einstein 1875–1948* (T-G 1969, 1982). A new edition appeared the following year with the same text but with a signed afterword (*Nachwort*) by her Swiss editor Werner G. Zimmermann and a more descriptive title, *Im Schatten Albert Einsteins: Das tragische Leben der Mileva Einstein-Marić* (In the Shadow of Albert Einstein: The

Tragic Life of Mileva Einstein-Marić) (T-G 1983). After the first volume of Einstein's *Collected Papers* was published, Paul Haupt issued a new edition of the book in 1988. In it the editor deleted her outdated statements in the light of the recently discovered Einstein-Marić correspondence and other biographical material and documents, and revised and amended parts of the text in an "Editor's Supplement." The 1988 edition also contained a new Afterword, written by the editor, containing fuller biographical information about the author (Zimmermann 1988a, 1988b). Amid the sudden interest in Marić arising from the publication of Einstein's papers and the subsequent articles by Troemel-Ploetz and Evan Harris Walker, the publisher issued the last edition of the book in 1993, referring to it as the fifth edition, although it was identical to the fourth (T-G 1988) edition. In 1995, a Belgrade publisher issued a reprint of the Serbian edition of Trbuhović-Gjurić's biography (T-G 1995).

Whether in Serbian, German, or French, Trbuhović-Gjurić laid the foundations of the Mileva Story in her biography. As Troemel-Ploetz put it: "She wanted to rescue Mileva Einstein-Marić from oblivion and write her into Serbian and scientific history" (T-P 1990, 417). That she did. Yet at the same time, she also introduced many errors, misconceptions, and exaggerations, owing in large part to her problematic use of sources.

FOUNDATIONS OF THE MILEVA STORY

In the Preface to her biography Trbuhović-Gjurić (1983 and 1988, 7) provided the following concise six-item summary of the Mileva Story and its foundations:

> From everything that we know about her [Mileva], we may conclude that her part in Albert Einstein's creative work was large and significant. We deduce this

1. from her great talent, already exhibited in her childhood;
2. from the power of her pursuit of knowledge and her penetration into mathematics and physics, which she undertook abroad despite the strong resistance of her surroundings, the prejudices of that time, and her small-town milieu;
3. from her unusual success in her studies;
4. from the collaboration with Einstein during the period when they were studying together and during their marriage;
5. from the statements by Albert Einstein himself about her participation in his work as confirmed by credible contemporaries;
6. from the fact that Einstein gave his Nobel Prize to Mileva.

These items, if corroborated by her book, would support Trbuhović-Gjurić's contention that "her part in Albert Einstein's creative work was large and significant" (T-G 1983 and 1988, 7). Certainly, we can agree with and admire number 2 in the list, "the power of her pursuit of knowledge." That aspiration led to her leaving her family and homeland for Switzerland; it underlay her determination to overcome the hurdles to women in education and science and to work towards acquiring a teaching diploma in physics and math as the only young woman in her class cohort at the Zurich Polytechnic. This achievement was certainly remarkable, and such determination would have been essential for keeping up with her scientifically ambitious classmate and future husband Albert Einstein and for, as Trbuhović-Gjurivić would have it, contributing to his work. But it does not necessarily mean that she did in fact contribute or that her contribution "was large and significant."

In this and subsequent chapters I examine the purported evidence for Marić's contribution more closely, considering the remaining five contentions presented by Trbuhović-Gjurić (numbers 1 and 3–6) along with some of the ways later authors have presented them.

TALENT: EARLY SCHOOLING AND GYMNASIUM (NO. 1)
Even before she started school, Marić's father Miloš taught his
talented daughter simple calculations and encouraged her to
learn German as an entrée into a future professional career (T-G
1983 and 1988, 21; Krstić 2004, 22).

In 1882, when she was six, Mileva entered a primary school
in the town of Ruma in the southern Hungarian province of
Vojvodina. According to Trbuhović-Gjurić, a teacher at the
school reportedly told her father: "Take good care of this child!
She is a rare ["seltsames"] phenomenon" (T-G 1983 and 1988,
21). That may have been the teacher's assessment, but we don't
know Trbuhović-Gjurić's source for this statement. It could not
have been Marić's father, since he died in 1922. So the statement, or some form of it, was likely handed down orally until it
reached Trbuhović-Gjurić while she was researching her book.

Nevertheless, while the quotation of the teacher's remark
remains hearsay, Dord Krstić (2004, 22), with his access to the
relevant documents, reports that in her last year in primary
school (1886) "her scholastic ratings were very good." In keeping
with Marić's performance in Ruma, Krstić states that at the end
of the single year that she spent in the first class at the Serbian
Higher Girls' School in Novi Sad (1886–87), "she received the
grade of 'excellent' in all subjects" (Krstić 2004, 22, n. 33; also,
T-G 1983 and 1988, 22). For the next three years she continued
her schooling at the "middle school" (Royal Lower Secondary
School) in Sremska Mitrovica. There, Krstić (2004, 23) reports,
"only Mileva among 14 pupils of her class, had excellent marks
in both mathematics and physics" at the end of her final year. He
adds that in all three years at the school "her general success was
the highest possible," but he cites no source for this claim.

Marić's next two academic years (1890–91 and 1891–92)
were spent at the Royal Serbian High School in Šabac.

Trbuhović-Gjurić waxes lyrical about Marić's time at this school, claiming that her exceptional talents already set her apart from the other students and that, in her isolation, she sought compensation "in a world of discoveries and possibilities that had been shaped over the centuries" (T-G 1983 and 1988, 24). Furthermore, "Mileva considered mathematics to be a great discovery of mankind that enabled the symbolic expression of all knowledge concerning nature and phenomena in general." Trbuhović-Gjurić again provides no source for these contentions: they are evidently fanciful notions based on her preconceptions about Marić. In addition, she states that during her time at the Šabac school Marić was the best in all subjects (T-G, 1983 and 1988, 24). Again, no source is given. Evidently this claim did not come from the school records, since she reports that these were destroyed, along with all the school equipment, in World War II (T-G 1983 and 1988, 26). Nevertheless, on the basis of her previous achievements it is evident from the reports that during her schooling in Vojvodina, Marić proved herself to be an excellent and conscientious student, including in mathematics and physics.

In 1892 Marić enrolled in the nearly all-male, state-sponsored Royal Upper High School in Zagreb. She was admitted as a private student, without the requirement to pay fees, and "even received financial support in the amount of 100 forints" (Krstić 2004, 29 and n. 51). There she remained for the sixth (repeated) and seventh of her eight gymnasium years. This time we do have her grade reports and other documentary material. (The Zagreb State Archive kindly provided me with her school records: see figures 1.1 and 1.2. The grade results are translated and summarized in appendix A.) According to the records, she was one of several private students that year. Her grades for both semesters during her first year (1892–93) at the high school were nearly all "very good" (corresponding to B on a scale A-E), including for mathematics. In the first semester, for Greek she obtained her only grade of "excellent" (A) while at the school. Although "Physics" is printed on the grade report, the grades for both semesters were left blank because, as the archivist Mihaela Barbarić wrote in the space, "They did not have Physics in 6th Grade" (figure 1.1).

It seems that something untoward must have occurred during Marić's second year at the Zagreb high school. For the first semester most of her grades dropped, and two were merely "satisfactory" (D). Her previous "very good" (B) in mathematics dropped to "good" (C), and her first-semester grade in physics was only "satisfactory" (D). (Another D was in German, which was B the previous semester.) There is no way to explain this sudden drop in performance without further information, but whatever it was she persevered and her grades partially recovered in the next semester. As shown in appendix A (from figure 1.2), these grades were all "good" (C), except for mathematics and physics which were "very good" (B).

Marić's grades of B for mathematics and physics were her best grades for that semester. Trbuhović-Gjurić, who cited another document from Marić's student records, presumably saw these grade reports, since she stated (ambiguously): "She had passed the final examination of the seventh class in September 1894 with the best grades in mathematics and physics" (T-G 1983 and 1988, 26–28). According to the grade report, these were actually *her* best grades for that final semester, and they were one grade below the highest possible. Dord Krstić (1991, 88) went further in his appendix to Elizabeth Roboz Einstein's memoir of her late husband Hans Albert Einstein, asserting that for Marić's 1894 examinations "she received the highest grades awarded in mathematics and physics," a claim repeated by Andrea Gabor

(1995, 5) and Radmila Milentijević (2015, 28). Peter Frize (2009, 274) went so far as to contend that she "showed she belonged [in physics class] by achieving the highest grade ever awarded by the school in mathematics and physics." But by 2004 Krstić had already revised his description, perhaps after reviewing her actual grade reports. In his book *Mileva & Albert Einstein* he wrote simply: "her best grades were in mathematics and in physics" (Krstić 2004, 30). From these grade reports, Marić was evidently a very competent and hardworking student. However, she did not achieve exceptional success, at least at this stage in her schooling, as Trbuhović-Gjurić claims.

Marić completed her schooling at the more congenial Zurich Higher Girls' School, which she attended from November 1894 to spring 1896. No grades for this period are cited by either Trbuhović-Gjurić or Krstić. That spring Marić passed the pre-university Matura examinations, but for some reason she was required to take the Zurich Polytechnic's mathematics entrance examinations, in which her grade average was 4.25 on a scale 1 to 6 (T-G, 1983, 56; 1988, 60; see appendix A). Again, these grades indicate that she was certainly competent in pre-university mathematics, but not that she was exceptionally talented as has been widely claimed.

SUCCESS IN STUDIES: POLYTECHNIC AND
DIPLOMA EXAMS (NO. 3)

Marić and Einstein entered the Zurich Polytechnic in fall 1896 (winter semester 1896–97) and enrolled, as the only two first-year physics students, along with four others specializing in mathematics, in the four-year program for future teachers of mathematics and physics. Half way through the program they had to pass an intermediate examination in order to continue. At the end of the four years they had to achieve an overall passing grade on the final exams combined with a graded research thesis in order to receive a diploma (Diplom), a degree comparable to a master's degree in the United States (*CPAE* 1, pp. 43–44). (Students generally entered higher education from their gymnasium [high school] at the level of the American junior year.)

Fortunately, the Zurich Polytechnic—renamed the Swiss Federal Institute of Technology, or ETH, in 1911—preserved in its archives all the semester grade records and exam grades for both Marić and Einstein. Trbuhović-Gjurić's research also led her to these archival records. She included in her book facsimiles of two pages from Marić's student grade report (Matrikel) (T-G 1983, 56-57; 1988, 60–61), along with a comparative table of course grades for Marić and Einstein (T-G 1983 and 1988, 43). In 1987 the archival grade records for Einstein appeared in the first volume of his *Collected Papers* (*CPAE* 1, doc. 28). The ETH has since placed Marić's grade report online (Marić Student Record, see bibliography). They all agree with the grades published by Trbuhović-Gjurić and are summarized in appendix C of this book.

A review of the course grades for both Marić and Einstein reveals that neither was a particularly outstanding student. Marić's grades were modest, averaging 4.7 on a scale 1-6. They do not indicate "unusual success in her studies" at the Polytechnic. Nor do they support Trbuhović-Gjurić's statement: "In Mileva he [Einstein] had found a serious, coequal comrade who was at times, even in mathematics, above him" (T-G 1983 and 1988, 44).

In her book Trbuhović-Gjurić did not discuss the two students' performances on the intermediate and final diploma exams. Apparently, she did not find those documents in the ETH archive. Einstein's exam grades were also published in

1987 (*CPAE* 1, docs. 42, 67) along with those of the other students in their small group, with the exception of Marić's intermediate exam results (see below). However, the latter were
recorded by Zimmermann (1988a, 63), the former Zurich city
archivist, in an editorial supplement to the 1988 edition of
Trbuhović-Gjurić's book. Marić's final diploma exam grades for
1901 were published along with Einstein's (*CPAE* 1, doc. 67),
but those for the 1901 final diploma exam (see below) are listed
only in an unpublished archival report submitted to the faculty
by Hermann Minkowski (Minkowski 1901, see document list
in bibliography), though Zimmermann had published her overall grade average in 1988 (Zimmermann 1988, 74).

Because she spent the first semester of the 1897–98 academic
year at the University of Heidelberg, Marić was permitted to
take the intermediate examination after her third year, in early
October 1899, rather than after her second year when Einstein
and the math students in their class took it. Both passed com-
fortably. According to Trbuhović-Gjurić's editor, who did find
the exam reports, Marić achieved a very creditable grade average
of 5.05 on a scale 1-6 (Zimmermann 1988a, 63). A year earlier
Einstein had achieved a grade average of 5.7, which was the
highest among the five students who took the intermediate
exams in 1898 (*CPAE* 1, doc. 42). While his grade of 5.5 in
physics was the same as Marić's, on all other topics he achieved a
higher grade (appendix D).

Trbuhović-Gjurić was clearly unaware of what happened
regarding the final diploma exams in July 1900. "Why she
[Marić] did not take the Diploma exam as he did, despite her
equal performance during the last two years of study, is not
known" (T-G 1983, 58; 1988, deleted). In fact, Marić did take
the final diploma exams in 1900 along with Einstein (appendix
D). Their grades in theoretical physics were respectively 4.5 and

5 (scale 1 to 6), and in the mathematics component (theory of functions), 2.5 and 5.5. They both had comparatively low grades (4 and 4.5 respectively) for the heavily weighted research thesis. The overall (weighted) averages resulted in a score of 4.91 for Einstein and 4.00 for Marić. Einstein's four subject grades were all 5 or above.

On July 28, 1900, Einstein was awarded his diploma (*CPAE* 1, p. 375). He was now certified as a Specialized Teacher in Mathematical Instruction (*Fachlehrer in mathematischer Richtung*) (*CPAE* 1, doc. 28, p. 50). Unfortunately, Mileva failed. The disappointing overall score of 4.00 prompted the math-physics section head, on behalf of the Committee of Examiners (all of the section professors), to recommend "not to award the Diploma to Fräulein Marić" (*CPAE* 1, doc. 67). It is not known how the committee reached this conclusion. No minimum passing grade or other performance requirements are stipulated in the ETH records or grade reports (Stachel 2002, 32). However, it is likely that Marić's very poor grade of 2.5 in the mathematics component was a major factor.

Gabor (1995, 15) has suggested, invoking Robert Schulmann (no citation given), that "because the finals included an oral component, she might have been subject to the prejudice of her examiners." But this did not hinder her from attaining good grades in the intermediate exams. (The issue of possible prejudice against Marić as a woman on the part of the examiners will be discussed in more detail in chapter 9.)

Trbuhović-Gjurić was also unaware that Marić retook the diploma exams the following year, in July 1901, while working toward her doctorate at the same time. Nor did she, or anyone else until 1986 with the belated discovery of the "love letters," know that Marić was nearly three months pregnant at the time and with no wedding in sight. Her thesis grade remained 4.00 as before, which meant that, amid her personal distress, she would have to perform even better on the subject exams. She did slightly better on two of the subject exams (increasing her mathematics grade to 3.5), but slightly worse on two others (see appendix D). The results were again an overall average of 4.00, which once again prompted the examining committee, headed by mathematics professor Hermann Minkowski, to recommend that Marić not be granted the diploma (Minkowski 1901).

Unaware of all this, Trbuhović-Gjurić attributed Marić's departure without a diploma to a dispute with the senior physics professor Heinrich Friedrich Weber: "She went so far to eventually withdraw her excellent Diplomarbeit [diploma thesis], stopped her research with him [Weber], and in August, 1901, left the Polytechnikum for good." (T-G 1983, 59; 1988, deleted; T-P 1990, 425). 78). How Trbuhović-Gjurić could know that Marić's diploma thesis was "excellent" is a mystery, since the Polytechnic did not retain them (*CPAE* 1, doc. 68, n.3). Her grade for the thesis submitted in 1900 was 4 out of 6. It is unlikely she submitted a second thesis.

A devastated Mileva, pregnant and unmarried, left the Polytechnic for her parents' home in Novi Sad, Vojvodina, without a teaching diploma.

ALBERT EINSTEIN: ON HER PARTICIPATION
IN HIS WORK (NO. 5)

For many readers one of the most convincing pieces of evidence that Marić participated in Einstein's scientific work comes from a hearsay report, recorded by Trbuhović-Gjurić, in which he is quoted as having told a group of "young intellectuals" in Novi Sad in 1905: "I need my wife. She solves all mathematical problems for me'" (T-G 1983, 75; 1988, 93).

In the introductory synopsis of her 1990 paper Troemel-Plo-etz (1990, 415) reproduced the second sentence in the above quotation: "My wife solves all my mathematical problems for me." That statement inspired the subtitle of her paper, "Mileva Einstein-Marić: The Woman Who Did Einstein's Mathematics," and the theme was reiterated in greater detail within the paper itself (T-P 1990, 421–422). (Elsewhere in the paper she quoted Trbuhović-Gjurić's above quotation in full [T.P. 1990, 418].)

Examining the context and source of this statement, we find that, according to Trbuhović-Gjurić, Einstein reportedly said these words at a gathering of student friends of Marić's brother Miloš Marić (a medical student at the time) during Einstein's first visit to Novi Sad in 1905 to be introduced to Marić's parents. It is possible to imagine Einstein making such a quip to an appreciative audience, but first we should consider the source of the report. It is attributed to a Dr. Ljubomir-Bata Dumić, one of the students present that day, but Trbuhović-Gjurić provides no other information about him. She also quotes him as having written:

> We raised our eyes toward Mileva as to a divinity, so much did her mathematical knowledge and genius impress us. Simpler mathematical problems she solved instantly in her head, and those which would have taken competent specialists several weeks of work she completed in two days. And always she found an original, unique way, the shortest. We knew that she had made him (Albert), that she was the creator of his fame. She solved for him all mathematical problems, especially those concerning the theory of relativity. It was simply amazing what a brilliant mathematician she was. (T-G 1983, 75; 1988, 93)

Reminiscences by an admiring acquaintance (who couldn't possibly have seen firsthand Marić's supposedly prodigious facility in mathematics), reported to Trbuhović-Gjurić some six decades after the event, can hardly be regarded as reliable. This is especially the case given what we know of her very modest prowess in mathematical topics at Zurich Polytechnic (see chapter 5). In any case, the mathematics in the 1905 paper on the theory of relativity to which Dumić alludes would not have stretched the expertise of a competent physics student, and it would have posed no difficulties for Einstein (*CPAE* 2, doc. 23). As the physics historian Jürgen Renn has observed, "If he had needed help with that kind of mathematics, he would have ended there" (Highfield and Carter 1993, 114–115). Likewise, the physicist Jeremy Bernstein (1998) writes, "There is no mathematics in these [1905] papers that would be beyond a suitably oriented university student. What is difficult are the ideas."

EINSTEIN'S ALLEGED MATHEMATICAL LIMITATIONS
Troemel-Ploetz attempts to approach the matter from the opposite direction by contending that Einstein's ability in mathematics was very limited, which would explain why he supposedly needed Marić's help in solving mathematical problems in his theoretical work. She supports this contention by "look[ing] at some self-evaluations of Albert Einstein before he had to play the role of genius of the century" (T-P 1990, 421). She then quotes the following passage in her English translation from the original German passage, which Trbuhović-Gjurić quotes from Einstein's "Autobiographical Sketch" of 1956:

> Higher mathematics didn't interest me in my years of studying. I wrongly assumed that this was such a wide area that one could easily waste one's energy in a far-off province. Also, I thought in my innocence that it was sufficient for the physicist to have clearly understood the elementary mathematical concepts and to have them ready for application while the rest consisted of unfruitful subtleties

for the physicist, an error which I noticed only later [with regret (left out of translation)]. My mathematical ability [*Begabung*] was apparently not sufficient to enable me to differentiate the central and fundamental concepts from those that were peripheral and unimportant. (T-P 1990, 421, quoting T-G 1983, 47, who quotes, without citing the source, from Einstein 1956a, 11)

Now *Begabung* is more accurately translated as "talent" not "ability." Einstein was not alluding to his ability to handle conventional mathematics, but to a more profound affinity with mathematics. This was expressed more fully in his more extensive intellectual autobiography written a few years earlier in 1949. There he explains how it was that he "neglected mathematics to a certain extent" when he was a student at the Zurich Polytechnic: "I saw that mathematics was split into numerous specialties, each of which could absorb the short lifetime granted to all of us. Consequently, I saw myself in the position of Buridan's ass, which was unable to decide upon any particular bundle of hay. Presumably this was because my intuition was not strong enough in the field of mathematics to differentiate clearly the fundamentally important, that which is really basic, from the rest of the more or less dispensable erudition." (Einstein 1949, 15)

On this topic, Troemel-Ploetz quotes a passage from Trbuhović-Gjurić in the context of the latter's depreciating Einstein's mathematical abilities and his consequently requiring help in that direction, seemingly retreating from her claim that Marić's "part in Albert Einstein's creative work was large and significant":

> In her work, she was not the co-creator of his ideas, something no one else could have been, but she did examine all of his ideas, then discussed with him and gave mathematical expression to his ideas

about the extension of Planck's quantum theory and about the special theory of relativity. (T-P 1990, 420; quoting T-G 1983, 72, T-G 1988, 90)

Unsurprisingly, given Marić's limitations in mathematics and the magnitude of her alleged contribution here to two major theoretical conceptions, the above claim of her mathematical role in these is entirely without foundation. It has long been established that the mathematical components of Einstein's extension of Planck's ideas in his 1905 paper on the quantum theory derived from developments in his three preceding fundamental papers on statistical mechanics and thermodynamics (Klein 1963; Jungnickel and McCormmach 1986, 248; *CPAE* 2, ed. note, 134–148). In addition, as already noted, the mathematical content of special relativity did not require any knowledge of the subject beyond that of any competent physics student.

Einstein's mathematical capabilities were more than adequate for the many other theoretical problems he faced during most of the first decade following completion of his Polytechnic studies in 1900. This is exemplified by the comments of Zurich University professor of physics Alfred Kleiner in his "Expert Opinion" on Einstein's doctoral dissertation of 1905, which Einstein successfully submitted to that university. Kleiner observed in his report: "The arguments and calculations to be carried out are among the most difficult ones in hydrodynamics, and only a person possessing perspicacity and training in the handling of mathematical and physical problems could dare to tackle them" (*CPAE* 5, Engl., doc. 31). Such were these difficulties that Kleiner recommended seeking the views of his colleague Heinrich Burkhardt, a professor of mathematics. Burkhardt reported in an addendum that he found Einstein's calculations to be "correct without exception, and the manner of treatment demonstrates *a thorough command of the mathematical methods involved*"

(*CPAE* 5, Engl., doc. 31, emphasis in original). (Nevertheless, it was later discovered that the dissertation did in fact contain a significant mathematical error, which Einstein corrected in 1911 [*CPAE* 2, ed. note, 179–182; *CPAE* 3, doc. 14].

HELP NEEDED WITH THE GENERAL THEORY According to Trbuhović-Gjurić (1983, 94, 96; 1988, 112, 114), by 1911 Einstein was no longer discussing his scientific work with Marić because of some discord that had arisen between them. Consequently, she writes, when he was occupied with the problem of gravitation and was having problems with the mathematics he no longer turned to his wife, but instead wrote to his mathematician friend Marcel Grossmann seeking his assistance: "I encountered mathematical difficulties which I cannot conquer. I beg for your help, as I am apparently going crazy" (trans. T-P 1990, 420–421).

Contrary to what Trbuhović-Gjurić states, and Troemel-Ploetz repeats, there is no such letter from Einstein to Marcel Grossmann, nor do we expect one since both were in Zurich at the time. The incident and quotation derive, not from a letter or document, but from the reminiscences of Einstein by a former mathematics student at the Zurich Polytechnic in Einstein's class, Louis Kollros. Kollros wrote that when Einstein returned to Zurich from Prague in summer 1912 as professor of theoretical physics at the ETH, he sought out Marcel Grossmann, also a former classmate and at that time a professor of mathematics at the ETH, and told him, "You must help me or else I'll go crazy!" (Kollros 1956, 27).

The background to this episode is that Einstein had been working in Prague in 1911-1912 on generalizing the theory of relativity to encompass accelerated motion, as in gravitation. This required a highly specialized mathematics known as differential geometry. As the physicist and historian Abraham Pais writes, in late March 1912 Einstein had written to another friend, Michele Besso: "Recently I have been working furiously on the gravitation problem. It has now reached a stage in which I am ready with the statics. I know nothing as yet about the dynamic field, that must follow next. ... Every step is devilishly difficult" (Pais 1982, 210; *CPAE* 5, doc. 377). Einstein recalled in 1923 that he made no further progress until, after returning to Zurich in 1912 to take up a professorship at ETH, he sought help from his colleague Grossmann, to whom he posed "the problem of looking for generally covariant tensors whose components depend only on the derivatives of the coefficients $[g_{\mu\nu}]$ of the quadratic fundamental invariant $[g_{\mu\nu}dx^{\mu}dx^{\nu}]$" (Pais 1982, 212). Evidently Einstein had a good idea of the geometry he required to make progress. He was already familiar with Carl Friedrich Gauss's theory of two-dimensional surfaces from Professor Carl Friedrich Geiser's Polytechnic course on differential geometry, and he realized he needed a generalization of it to higher dimensions. In response to Einstein's request for assistance, Grossmann consulted the literature and reported that there was indeed such a geometry, a non-Euclidean geometry involving curved spaces, but it was outside of Grossmann's current expertise (Pais 1982, 213; also, Renn et al. 2007).

Following Grossmann's suggestion, Einstein was able to write to the Munich theoretical physicist Arnold Sommerfeld in October 1912: "I am now working exclusively on the gravitation problem and believe that I can overcome all difficulties with the help of a mathematician friend of mine here." His additional remarks in this letter to Sommerfeld provide a further clarification of the words quoted by Troemel-Ploetz from Trbuhović-Gjurić (above), which both authors misconstrue as demonstrating Einstein's limited mathematical abilities, even in the early years:

Figure 6.1
Marcel Grossmann, 1909. Courtesy of ETH Library Zürich, Photo
Archive, portrait 01239, photographer unknown. Public domain mark.

But one thing is certain: never before in my life have I troubled
myself over anything so much, and I have gained enormous respect
for mathematics, whose more subtle parts I considered until now, in
my ignorance, as sheer luxury! Compared with this problem, the
original theory of relativity is child's play. (*CPAE* 5, doc. 421)

By 1913, Einstein and Grossmann had developed and published in that year an "outline of a generalized theory of relativity and of a theory of gravitation" (*CPAE* 4, Engl., doc. 13), in which Grossmann provided a mathematical section that served as a primer for anyone entering this field. Only a brief perusal of this paper will convince readers of the mathematical difficulties they encountered.

REPORT OF A FORMER STUDENT   Troemel-Ploetz, drawing again upon Trbuhović-Gjurić, offers further evidence of Einstein's alleged mathematical shortcomings—a hearsay report from a former student regarding an incident that occurred while Einstein was lecturing at the University of Zurich. She writes:

> A former student of Einstein recalls that Albert Einstein got stuck in the middle of a lecture missing a "silly mathematical transformation" which he couldn't figure out. Since none of the students could either, he told them to leave half a page empty and gave them the result. Ten minutes later he discovered a small piece of paper and put the transformation on the blackboard, remarking: "The main thing is the result not the mathematics, for with mathematics you can prove anything." (T-P 1990, 421)

This paragraph in Troemel-Ploetz's paper is taken almost verbatim from Trbuhović-Gjurić (1983, 87–88; 1988, 105). This time Trbuhović-Gjurić did state her source for this story but not where to find it. She identified a Dr. Hans Tanner as the former student in question. However, a lengthy passage from Tanner's recollections of Einstein's lectures at the University of Zurich during the years 1909–1911 appears in the documentary biography of Einstein by the Swiss author Carl Seelig (1954, 119–126). From this we can see that the account appearing in Trbuhović-Gjurić's book is a somewhat garbled version of what Tanner actually wrote, according to Seelig's translation:

During the whole time, as far as I can remember, Einstein got stuck only once. He suddenly stopped in the middle of a lecture and said: "There must be some silly mathematical transformation which I can't find for the moment. Does anyone see it?" We naturally did not see it. "Then leave a quarter page blank! We don't want to waste any time." Some ten minutes later Einstein interrupted himself in the middle of an elucidation. "I've got it." We had no idea what he had. During the complicated development of his theme he had still found time to reflect upon the nature of that particular mathematical transformation. That was typical of Einstein. (Tanner, quoted in Seelig 1956a, 101)

Contrary to Trbuhović-Gjurić, the "small piece of paper" supposedly consulted by Einstein to rescue the situation does not figure in this incident: Tanner mentioned it earlier in a *different* context (Seelig 1956a, 100). Likewise, the quotation of Einstein's words about proving anything with mathematics actually occurred in relation to an entirely different episode recounted in Seelig (1956a, 102–103). Furthermore, whereas Tanner has concluded his report in terms that reflect positively on Einstein's mathematical capabilities, Trbuhović-Gjurić concludes from her misleading account: "So this is how Einstein viewed mathematics: That others had to provide him [Einstein] the proofs" (T-G 1983, 88; 1988, 106). Troemel-Ploetz went a step further: "He did not have to worry about the proofs because Mileva Einstein-Marić was doing them" (T-P 1990, 421).

EINSTEIN'S NOBEL PRIZE: A MATTER OF MONEY OR
CREDIT (NO. 6)

Einstein did not simply "give his Nobel Prize [money] to Mileva," as Trbuhović-Gjurić stated (T-G 1983 and 1988, 8). But not until documentation regarding their divorce settlement in 1918 (finalized in 1919) and the award of the prize in 1922 (for

1921) was published in the later volumes of the *Einstein Papers*, in 1998 and 2004, has the full story become available.

Einstein received the Nobel Prize "for his services to theoretical physics, and especially for his discovery of the law of the photoelectric effect" (Nobel Archive, see bibliography). The law derived from his 1905 paper on the quantum theory of light, a paper that appeared while he was still married to Marić. By the 1980s it was generally understood that, as Abraham Pais (1982, 300) wrote, the divorce decree "stipulated that Mileva would receive, in due course, Einstein's Nobel prize money." In the biographical sketch of Mileva Einstein-Marić appearing in volume 1 of the *Collected Papers* in 1987, the editors stated: "1922, given Einstein's Nobel Prize money, in accordance with divorce agreement; later Einstein purchased residence and income property for her" (*CPAE* 1, p. 381).

Proponents of the Mileva Story and critics of Einstein reacted accordingly. Troemel-Ploetz asked: "How did it happen that she only got the money from the Nobel Prize and was not named a winner together with Einstein?" (T-P 1990, 416). Both she and Evan Harris Walker speculated that Einstein's sharing of the financial award with Marić was private payment for her unrecognized work. Walker (1989, 11), who wrote "I find it difficult to resist the conclusion that Mileva, justly or unjustly, saw this as her reward for the part she had played in developing the theory of relativity," overlooked the fact that the prize was awarded for the law of the photoelectric effect. Milan Popović wrote in 2003: "That Albert gave Mileva all of the money he received for winning the Nobel Prize in Physics for 1921 was seen by some as acknowledgment of his intellectual debt to her" (Popović, quoted in M-KS, 23).

Troemel-Ploetz explored the possibilities further:

When Albert Einstein received the Nobel Prize in 1922, he had been separated from his wife, and living with another woman in Berlin for eight years; he had been divorced and remarried for three years. However, he traveled to Zurich and gave the full financial award, which came with the Nobel Prize, to his first wife.

Many interpretations are possible, of course. People say he turned over the Nobel Prize to his wife. This is simply a harmonizing euphemism. *He* [in original] was the one who received the prize with all the honors, he did not renounce it in her favor, and it was he who gave the lecture in Göteborg at the congress of Nobel Prize winners. Perhaps he only gave the money to his first wife because for eight years he had hardly supported her and the two children at all financially.

The *Collected Papers of Einstein* [sic], *Vol. 1*, suggest a different reason. I was amazed to read there that Mileva Einstein-Marić was given the Nobel Prize money in accordance with the divorce agreement [ref in original: *CPAE* 1, 381]. I asked myself whether the divorce agreement of 1919 anticipated Einstein's Nobel Prize of 1922. But let us assume that he was giving her private recognition for her contribution which he had not given her publicly. By then, he must have been aware of how much he owed her mathematical genius; his own genius was on the decline and he did not achieve anything comparable to what is defined as his "creative outburst of 1905." Again and again people remarked on the fact that none of his later work, after the age of 26, surpassed or even reached the same level as his earlier research. (T-P 1990, 420)

This statement contains many inaccuracies. First of all, regarding Einstein's supposed scientific decline after 1905, when he was twenty-six, Einstein, far from being in decline, began work in 1907 on what would become one of the greatest achievements of twentieth-century science, the general theory of relativity! By the time of his separation from Marić in 1914, he had been working on it for seven years and published his first paper on the full-scale general theory in November 1915 (*CPAE* 6, doc. 21). So far it has passed every test of its validity. In 2016 scientists discovered the existence of gravitational waves, propagating distortions of space-time that Einstein had predicted exactly a hundred years later. The theory has been a solid foundation of contemporary physics since its inception.

Walker (1991, 122), however, claims that "Einstein's completion of the general theory of relativity shortly after his separation from Marić involved nothing but the conclusion of work that was already eight years in development." The implication is that Marić had helped him with the development of the general theory during those eight years, after which he simply completed what they had already mostly achieved together. But, as we have seen, the mathematics of the new theory was so difficult that Einstein himself was able to pursue the theory only with the help of his friend, the ETH mathematics professor Marcel Grossmann. Moreover, as can now be seen in Marić's letters to her friend Helene Kaufler Savić during those years, rather than working closely with her on general relativity, Einstein was increasingly distant from her and immersed alone in his science. This was one reason the marriage failed. For instance, in December 1912 she complained to Kaufler Savić, "He is tirelessly working on his problems, one can say that he lives only for them. I must confess with a bit of shame that we are unimportant to him and take second place. We are living here happily and fairly quietly, since my husband does not like interruptions to his work" (M-KS, 107–108; see also M-KS: Summer? 1909, 94; Winter 1909/1910, 102). One may add that the formidable intricacies of the researches that occupied Einstein in this period required a mastery of the basic elements of the subject that only a dedication of the kind Marić complains about in 1912 could accomplish. A year later, Einstein and Grossmann completed their "outline" of the new theory. Historians have shown that there was still much work remaining after the separation until Einstein's first paper on the full-scale theory in November 1915 (Janssen and Renn 2015; Renn et al. 2007).

Second, regarding Troemel-Ploetz's financial claims in the above extracted passage, it is not the case that, prior to 1922, Einstein "had hardly supported [his wife] and the two children at all financially." Documents published subsequent to Troemel-Ploetz's assertion in 1990 contradict her assertion. After Einstein and Marić separated in 1914, he lived alone in an apartment until his marriage to his cousin Elsa in 1919. In the separation agreement he committed himself to supporting Marić and the children with "at least" 5,600 German marks per annum in quarterly installments, which was just under half his annual salary of 12,000 marks at the time, and comparable to his annual Zurich salary in 1910–11 (*CPAE* 8A, doc. 40; Fölsing 1997, 328–329). In 1918 he raised the annual payment to 9,000 marks per annum, owing to the decline of the mark relative to the Swiss franc during the war.

Troemel-Ploetz's statement (in the long extract above) that Einstein "traveled to Zurich and gave the full financial award, which came with the Nobel Prize, to his first wife" is erroneous because the divorce decree stipulated that the prize money was to be deposited in a trust in a Swiss bank, and that Marić could dispose freely of the interest, but had no authority over the principal without Einstein's consent (Milentijević 2015, 420–421; see also below). So the claim by Krstić (1991, 98), that "Einstein accepted the prize check and gave it to Mileva," is also erroneous.

According to the later published documents, the divorce agreement of June 12, 1918 (in preparation for the final divorce decree in 1919) stated that Einstein would deposit (he already had) 40,000 German marks in a Swiss bank for the support of

Marić and the children, in addition to the payments from his salary. Upon their divorce it would become her property, but she could draw freely only upon the interest from the date of deposit. She could not withdraw from the principal without Einstein's approval (*CPAE* 8B, docs. 449 and 562). As the German mark continued to fall during and after the end of the war, and the family's medical expenses rose, the Nobel Prize money became a convenient source for maintaining his support of the family. The divorce settlement stated that, should Einstein later receive the Nobel Prize, he would deposit in a Swiss bank the entire prize money minus the 40,000 marks he had previously deposited. The total trust would thus amount to the entire prize award. Again, Marić would have free access only to the interest and would need to obtain his permission to access the principal. Upon her death or remarriage, the principal would go to their children. In the meantime, Einstein would set aside an additional 20,000 marks in a German bank. In the event that he died without receiving the Nobel Prize, the interest would go to her. After receiving the prize, rather than the annual payments of 9,000 marks, Einstein would ensure that Marić and the children received a minimum income of interest and payments of 8,000 Swiss francs (rather than volatile marks) per year. A Swiss court in Zurich finalized the entire arrangement upon issuance of the final divorce decree on February 14, 1919 (*CPAE* 9, doc. 6; *CPAE* 8B, doc. 562; Fölsing 1997, 420–421; Isaacson 2007, chap. 10).

According to Walter Isaacson, who examined archival material newly released in 2006, Einstein did provide Marić all of the funds he was obliged to pay. He also covered Eduard's and Mileva's medical costs, and he subsidized her living expenses.[2] Einstein received the Nobel Prize in 1922, with a financial award equivalent to about 180,000 Swiss francs (Fölsing 1997, 542).

Marić requested a quarter of those funds as a down payment for the purchase of a multifamily rental property and residence for her and the boys. Einstein agreed and sent her the money, but he wanted to place the rest of the money with a New York investment bank, rather than in a Swiss bank per the divorce agreement (*CPAE* 14, doc. 37). Following an argument over this, Einstein placed another 45,000 francs in a Swiss bank under her name, as required, but, in violation of the agreement, he invested the rest with the New York investment firm under her name (*CPAE* 14, docs. 204 and 205; Wolff 2016; Milentijević 2015, chap. 18).

There is no indication in the available documents that either Marić or Einstein saw the invested Nobel Prize money in her name as payment for her contributions to his work. Nor did it even come up, for which we turn to Heinrich Medicus. Medicus was in 1918 a young Zurich physicist (doctorate the same year). Later he received access to the private correspondence of Einstein's close friends Heinrich Zangger, Michele Angelo Besso, and Besso's wife Anna Besso, née Winteler. All three were closely involved in advising Einstein and Marić as difficulties arose between them over the separation and divorce agreements. Medicus writes:

> It would have been conceivable that in the letters between the Bessos and Zangger the contributions of Mileva to Einstein's early work in relativity—whatever they may have been—might have been mentioned, perhaps as bargaining points in the negotiations for her financial support. However, there is nothing in the correspondence, even in connection with the Nobel Prize, that could indicate her scientific help. (Medicus 1994, 470)

In her biography of Mileva Einstein-Marić, Trbuhović-Gjurić listed six foundations for her conclusion that Marić's part in Albert Einstein's creative work was extensive and significant. In chapter 6, I looked at five of these assertions. The remaining one (number 4) alludes to Marić's alleged collaboration with Einstein during their student days and during their marriage. In this chapter I examine the evidence pertaining to the period when they were studying on the same course at Zurich Polytechnic. In chapter 8, I will turn to the alleged collaboration during their marriage.

Following the publication of the first Serbian and German editions of Trbuhović-Gjurić's biography in 1969 and 1982, new and more reliable documentary evidence concerning the scientific and personal relationships between Marić and Einstein began to appear, especially regarding the early years before their marriage in 1903. Although this evidence was not available to Trbuhović-Gjurić, it was available to Troemel-Ploetz, Walker, and all subsequent authors. I focus here on this new evidence, which, even though more reliable than hearsay, still became the subject of controversy and a new alleged underpinning of the Mileva Story.

By new evidence I refer to the publication in 1987 of the first (of currently fifteen) volumes of *The Collected Papers of Albert Einstein* (*CPAE* 1). Included in that first volume were fifty-one

of fifty-four letters, rediscovered in 1986, exchanged between
Marić and Einstein from October 1897 to February 1902, the
years during and just after their studies at the Zurich Polytechnic. (The dates of the remaining three fell outside of the remit
for volume 1, and were subsequently published in 1995 in volume 5.) In the case of the letters that were undated, the editors
assigned approximate dates or date ranges to each based upon
their contents. They had been in the possession of the family of
Marić and Einstein's son Hans Albert Einstein. He had come to
the United States in 1938 and became an engineering professor
at the University of California, Berkeley. The letters had come to
the attention of the Einstein editorial project years earlier, but
their location remained unknown. However, as a consequence
of persistent sleuthing by the historian Robert Schulmann,
Hans Albert's daughter Evelyn Einstein came upon the letters
among papers in her possession in Berkeley in 1986. As noted
above, all but three of them were published the following year in
the *Collected Papers* in the original German and in the English
translation volume published in the same year. A more polished
translation of the letters through September 1903 appeared in
1992 (E-M). The publication of these letters, wrote Highfield
and Carter (1993, 278–281), "revolutionized our understanding of Einstein's early years."

At that time, a second trove of letters also became available to
the Einstein Papers Project. These were the letters from Marić to
her close friend Helene Kaufler Savić from 1899 to 1932, both
before and after Helene's marriage. As far as is known, none of
the letters from Kaufler Savić to Marić have survived. Kaufler
had come to Zurich from Vienna in order to study history at
Zurich University. As Marić also experienced, Austria-Hungary
did not yet admit women to higher education. Kaufler married
the Serbian engineer Milivoj Savić shortly after graduating in
November 1900, and they were soon settled in Belgrade, Serbia,

with their two daughters. Volume 1 of the *Collected Papers* also contained brief excerpts from the early Marić–Kaufler Savić letters that referred to Einstein, as well as notes that Einstein had written at the end of Marić's letters to her. But the letters themselves were not published until 1998 when Kaufler Savić's grandson Milan Popović published the complete letters in their original languages, mostly German, with a parallel Serbian translation (Popović 1998). In 2003 he provided an English translation, but without the original languages (M-KS).

THE EINSTEIN-MARIĆ LETTERS

The Einstein-Marić letters naturally held special interest to those pursuing the biography of Mileva Einstein-Marić. Of the fifty-one letters in volume 1 of the *Collected Papers*, only eleven are from Marić, while forty are from Einstein. Many of the early Marić letters were evidently discarded by Einstein; he was not particularly concerned at that time and for some years after about saving any of the letters he received (or the manuscripts he wrote). In his letter to Marić dated December 17, 1901, he wrote, while discussing his living quarters, "You know what a dreadful state my worldly possessions are in—it's lucky I don't have much" (E-M, 69).

The most striking claim made by some authors who read the Einstein-Marić correspondence was germane to the emerging Mileva Story. The claim was articulated most comprehensively by Evan Harris Walker in a paper he presented to the AAAS meeting in 1990 and, more concisely, in a letter published in *Physics Today* in 1991:

> I find statements in 13 of his 43 letters to her [note 6] that refer to her research or to an ongoing collaborative effort—for example, in document 74 [in *CPAE* 1], "another method which has similarities

to yours." In document 75, Albert writes: "I am also looking forward very much to our new work. You must continue with your investigation." In document 79, he says, "we will send it to Wiedemann's *Annalen* [*der Physik*]." In document 96, he refers to "our investigations"; in document 101, to "our theory of molecular forces." In document 107, he tells her: "Prof. Weber is very nice to me ... I gave him our paper." (Note 6: Documents 50, 57, 74, 75, 79, 93, 94, 96, 101, 102, 107, 111, and 127 [Walker 1991, 122 and 124])

EVALUATING THE LETTERS

Walker's argument that Einstein himself referred in writing to "our investigations," "our theory," "our paper," and "our work on the relative motion" (*CPAE* 1, doc. 94) does, on the surface, suggest that they engaged in collaborative work related to Einstein's extracurricular researches on advanced physics not covered in their course at the Polytechnic. Such written material must be taken seriously, considerably more so than hearsay or memory. Yet it should be noted that none of Einstein's extant letters to her contain a response to any conceptual or mathematical contribution that she may have made in a (now-lost) letter to him. Such was his enthusiasm about the flood of ideas he was producing, and the books by eminent physicists that he mentions he was reading, that it is inconceivable that he would have made no comment in reply to any suggestion coming from Marić had there been one. This does not preclude the notion of her having played a collaborative role, but it does suggest that Walker's contentions should be examined on a case-by-case basis. One should keep in mind that we can have no knowledge of any ideas exchanged when they were both in Zurich, though much of the time from July 1900 to September 1902 they lived in different places (Krstić 2004, 63, 66, 70–71, 88–89). It is certainly possible that during the periods when they were both in Zurich, Marić made suggestions to Einstein in relation to his researches on which he was always eager to communicate, for example, with his former classmate Marcel Grossmann in April and September 1901 (CPAE 1, docs. 100 and 22, respectively). But it should also be noted that Marić could hardly have worked on any of Einstein's current preoccupations on her own in the last months of 1901 to early summer 1902, as she suffered through a difficult latter stage of pregnancy and the traumatic birth of their baby Lieserl in late January 1902 (Zackheim 2000, 41–43). She spent this whole period at her parents' home in Novi Sad.

It has been argued (e.g., Stachel 2005, li–lii) that the broader personal and emotional context of the words highlighted by Walker above must be taken into account in attempting to grasp their significance. Einstein wrote the letters containing plural possessive pronouns almost entirely during the period from March 1901 to December 1901, with the exception of one dated September 13?, 1900, that alludes to their respective doctoral dissertations (see below), and another dated October 3, 1900 which is in relation to results Einstein had obtained on capillarity. The twelve months from mid-1901 to mid-1902 were especially difficult for Marić—she was trying to come to terms with her two failures in the Polytechnic final diploma exams and the dashing of her ambition for a career in science. As if this wasn't enough, she was pregnant, and her giving birth to baby Lieserl in late January 1902 was traumatic (Zackheim 2000, 40–42). On top of all that, as she told Kaufler Savić, she experienced anguish during the lengthy periods of separation from Einstein with no immediate prospect of marriage (M-KS, 78). Einstein's concern for her at this time, it has been argued (e.g., by Pais [1994, 8]), led to his using expressions such as "our work" and "our paper" in order to help her emotionally by

drawing her into his own researches. Highfield and Carter
(1993, make a similar argument in relation a much-quoted sentence in a letter from Einstein to Marić in March 1901 in which
occurs the phrase "our work on relative motion." They quote the
relevant passage to make their point, adding: "By italicizing the
key sentence, one shows how it sat marooned, not in one of
Einstein's many passages of close scientific argument, but amid
an outpouring of reassurance that his love for Mileva remained
absolute despite their separation."

Regardless of whether Highfield and Carter's suggestion has
any substance, it is pertinent to note Einstein's fond hope,
expressed in his letters to Marić, for a future life in which the
couple would work together on science (E-M, 52, 71, 73). With
the above in mind, Einstein's use of first person plural pronouns
should be examined more closely than most commentators have
done hitherto.

EXAMINING QUOTATIONS FROM THE LETTERS

In this section I examine the quotations cited by Walker (above)
and the conclusions he and others drew from them. Walker used
the English translations available to him in the English language
version of *CPAE*, volume 1, with some slight corrections. The
more polished translations by Shawn Smith did not appear until
1992 (E-M). I begin with Walker's first two quotations:

"*another method which has similarities to yours*" (doc. 74), and "*I am
also looking forward very much to our new work. You must now continue with your investigation*" (doc. 75). The contexts in which these
two quotations (italicized) appear are as follows:

For the investigation of the Thomson effect I have again resorted to
another method, which has some similarities with yours for the determination of the dependence of κ on T & which indeed presupposes

such an investigation. If only we could already start tomorrow! With Weber we must try to get on good terms at all costs, because his laboratory is the best and the best equipped. (*CPAE* 1, Engl., doc. 74)

I am also looking forward very much to our new works [Arbeiten]. You must now continue with your investigation—how proud I will be when maybe I'll have a little doctor for a sweetheart while I am myself still a totally ordinary man. (*CPAE* 1, Engl., doc. 75)

Both of these letters are from the period from August to September 1900, after Einstein and Marić had submitted their theses in July 1900 as required for the diploma. Both chose research topics on heat conduction for their theses (*CPAE* 1, doc. 74, n. 6; doc. 75, n. 5; M-KS, 60, 67). On July 14, 1900, Marić's friend Helene Kaufler wrote to her mother: "Miss Marić and Mr. Einstein have now finished their written examinations [and theses]. They devised their topics together, but Mr. Einstein left the nicer one to Miss Marić" (M-KS, 60).

After Einstein graduated in July 1900 he planned to use Weber's laboratory for doctoral research on electrochemistry, with Weber as his dissertation adviser (*CPAE* 1, docs. 74, 82; Fölsing 1997, 74). At the same time, following her failure to pass the diploma exams, Marić intended to take the exams again the following year, while working under Professor Weber's supervision to expand her thesis research into a doctoral dissertation. Einstein encouraged her, as quoted above: "You must now continue with your investigation—how proud I will be when maybe I'll have a little doctor for a sweetheart while I am myself still a totally ordinary man." It was in this context that he wrote: "I am also looking forward very much to our new works" (i.e., their respective dissertations). In the other letter containing Walker's quotation he explained that "for the investigation of the Thomson effect I have again resorted to another method, which has some similarities with yours for the determination of the dependence of κ on T ..." (i.e., in her dissertation research). Marić alluded to their respective dissertation plans when she wrote to Helene Kaufler from Zurich in October 1900: "For the time being I am studying at home with Albert; next week we begin our laboratory work" (M-KS, 67); and in December 1900 she wrote: "Albert is still here [in Zurich] and will remain here until he finishes his doctoral dissertation, which will probably take until Easter" (M-KS, 69). In the fall of 1901 Marić (M-KS, 78) reported to Kaufler-Savić: "I have finished my studies, although, thanks to Weber's concerns, I have not yet managed to obtain a doctorate."

It is evident that the first two quotations cited by Walker allude, not to Einstein's (or indeed Marić's) extracurricular researches as Walker implies in his letter to *Physics Today* (Walker 1991, 122), but to the couple's respective Polytechnic-based doctoral dissertations. There is nothing in the two passages in question (also E-M, 30, 32) to indicate that they were *collaborating* on each other's doctoral dissertations, though no doubt they would occasionally have discussed together aspects of their respective researches. Nor does Marić's project mean that she was involved with the kind of extracurricular explorations implied by Walker when he writes of "her research [and her] ongoing collaborative effort" (Walker 1991, 122). Despite Einstein's inclusive use of "our new works" in the earlier letter, it is clear that the doctoral dissertation in question, now extended to include gases, was Einstein's alone, and there is no evidence of any contribution from Marić. This is also evident from the following passage in a letter of Marić's sent to Helene Kaufler Savić (now married) in December 1901: "Albert has written a magnificent study, which he has submitted as his dissertation. I have read this work with great joy and real admiration or my little darling, who has such a clever head" (M-KS, 79–80).

Marić's writing that she has "read this work with great joy" indicates that she could scarcely have been involved in its production. As for Einstein, he had stopped working with Weber on his dissertation in early summer 1901. After working independently, he submitted it to the Zurich University professor of Experimental Physics Alfred Kleiner in November 1901, but withdrew it after several months (*CPAE* 1, docs. 125, 132).

"we will send it to Wiedemann's Annalen" (*CPAE* 1, doc. 79)*; and "our investigations"* (*CPAE* 1, doc. 96)

In his *Physics Today* paragraph (Walker 1991), Walker follows the quotation "You must continue with your investigation" (from doc. 75) with "we will send it to Wiedemann's *Annalen*" (from doc. 79), suggesting, no doubt unintentionally, that "it" refers to Marić's investigation (her ongoing doctoral research on heat conduction). However, the quotations refer to different works, and were written a month apart. The full context of the first quotation immediately above, from a letter dated October 3, 1900, is follows:

> The results on capillarity, which I recently found in Zurich, seem to be totally new despite their simplicity. When we're back in Zurich, we shall seek to get empirical material on the subject through Kleiner. If a law of nature emerges from this, *we will send it to Wiedemann's Annalen.* (*CPAE* 1, Engl., doc. 79)

Einstein was by then intensively researching capillarity on his own, and "it" refers to a law emerging from the results that *he* had recently found in Zurich. (See below for more details on the subject of capillarity.)

Walker's second quotation above, alluding to "our investigations," comes in a passage within a lengthy paragraph in a letter dated April 4, 1901:

> Michele [Besso] arrived with wife and child from Trieste the day before yesterday. ... Yesterday evening I talked shop with him with great interest for almost 4 hours. We talked about the fundamental separation of luminiferous ether and matter, the definition of absolute rest, molecular forces, surface phenomena, dissociation. He is very interested in *our investigations*, even though he often misses the overall picture because of petty considerations. ... The day before yesterday he went on my behalf to see his uncle Prof. Jung, one of the most influential professors of Italy & [I] also gave him *our paper*. (*CPAE* 1, Engl., doc. 96)

Einstein unspecific reference to "our investigations" follows immediately after his referring to topics on which he had evidently given some considerable thought, since he spent nearly four hours discussing them with Besso. There is no evidence that Marić had given much thought, if any, to these topics. The use of "our" in this context can either be taken literally, or be seen as an example of Einstein's striving to draw Marić into his wide-ranging theoretical researches. However, it should be judged in the light of the other items discussed in this chapter, which provide no evidence of Marić's involvement with Einstein's researches.

At the end of the passage quoted above, Einstein wrote that he gave Professor Jung "our paper." The editors of volume 1 of the *Collected Papers* identify this as Einstein's paper on capillarity that appeared in *Annalen der Physik* a month before this letter (*CPAE* 1, doc. 96, n.8; *CPAE* 2, doc. 1, dated December 13, 1900; published March 1, 1901). One week after Einstein submitted the paper for publication, Marić wrote to Kaufler Savić:

Albert wrote a paper on physics that will probably soon be published in *Annalen der Physik*. You can imagine how proud I am of my darling. This is not an everyday kind of paper but is a very important one; it deals with the theory of liquids. (M-KS, 70)

Milentijević (2015, 69) explains Marić's assigning the paper to Einstein alone as exemplifying her "extreme modesty." Leaving aside that she provides no evidence that Marić *was* extremely modest, it is difficult to think of a reason why she would either hide any contribution she had made from her closest friend—to whom she had poured her heart out in her previous letter, in fall 1901 (M-KS, 76–77)—or express such pride in her fiancé's achievement if she had been his collaborator. A little earlier Milentijević had written about the period leading up to the completion of the paper by December 13, 1900 (*CPAE* 2, doc. 1): "After they returned to Zurich in October 1900, they had proceeded to work together in earnest on a project dealing with capillarity." However, she cites no evidence for this contention. In mid-September 1900, as noted above, from his parents' home in Milan Einstein wrote that he was looking forward to working on their new papers, i.e., their respective dissertations: in his previous letter (August 30 or September 6) he had referred to the different experimental methods which they would be using as part of their doctoral research.

Milentijević claims that there is evidence for Marić's "contribut[ing] to the capillarity project" from "Albert's own description in his October 3 letter to her" (Milentijević 2015, 69). This rather overstates what is in the letter. Writing again from Milan, Einstein told Marić: "The results on capillarity which *I* recently obtained in Zurich seem to be entirely new despite their simplicity" (E-M, 36, emphasis added). He continues: "When we're back in Zurich we'll try to get some empirical data on this subject from Kleiner." But, again, *Einstein* is the one who (eventually) made contact with Kleiner (E-M, 62, 63, 66, 69). There is not a single piece of evidence that Marić worked on the capillarity project with Einstein. But Einstein's writing from Milan of his results on capillarity "which [he] recently obtained in Zurich" and which "seem to be entirely new" is an indication that he was sometimes working on his research on capillarity when he was away from Zurich, a period that lasted from late July to mid-September 1900 (E-M, 19, 21, 22, 27, 29, 31) apart from his brief visit to Zurich mentioned above (E-M, 24, 25).

Milentijević (2015, 69) acknowledges that "[t]here is no doubt that Albert played the leading role in the [capillarity] project." And yet, she writes, "There is also little doubt that the paper was the result of the joint labor of two physicists working in collaboration. A scientific paper resulting from a partnership in today's world would have been authored by both scientists, or, at a minimum, each would have received credit for his or her contribution." From the above examination of Milentijević's claims, however, it is evident that there is no documentation of any specific contribution made by Marić to the capillarity paper, and her attributing it solely to Einstein while expressing pride in his achievement suggests that any (unidentified) assistance she might have given could only have been marginal.

"*our theory of molecular forces*" (doc. 101)

This quotation occurs in the following context, from a letter written by Einstein to Marić in Zurich while he was visiting his family in Milan.

> As for science, I've got an extremely lucky idea, which will make it possible to apply *our theory of molecular forces* to gases as well. You certainly remember that the force function appears explicitly in the integrals that have to be evaluated for the calculation of diffusion,

thermal conduction & viscosity. Hence, with gas molecules, *only* our constants c_α are necessary for the calculation of these coefficients for ideal gases, and one does not have to venture into the theoretically so uncertain area of deviations from the ideal gas state. I can hardly await the outcome of this investigation. If it leads to something, we will know almost as much about the molecular forces as about the gravitational forces, and only the law of the radius will still remain unknown. (*CPAE* 1, Engl., doc. 101)

In the letter above, dated April 15, 1901, as well as in a letter from the previous September (E-M, 32), when Einstein mentioned molecular forces in relation to his dissertation he discussed only his *own* ideas. Writing to his friend Marcel Grossmann just a day earlier, on April 14, he had mentioned the notion expressed above—to apply a theory used in the capillarity paper (i.e., *our theory of molecular forces*) to extend beyond liquids to gases. But in that letter, he did not mention Marić:

As for science, I have a few splendid ideas, which now only need proper incubation. I am now convinced that *my theory of atomic attractive forces* [emphasis added] can also be extended to gases, and that it will be possible to obtain the characteristic constants of almost all elements without great difficulty. (*CPAE* 1, Engl., doc. 100)

Grossmann must have been aware that the couple had frequently studied together, so what harm would there have been for Einstein to write him something like "my theory ... which I developed with some help from Miss Marić" if such had been the case? (See also below in relation to another letter [*CPAE* 1, doc. 102] to Marić in which Einstein again uses the inclusive "our" in relation to molecular attractive forces, which evokes no response in her letter in reply.)

"Prof. Weber is very nice to me ... I gave him our paper" (doc. 107)

This last quotation provided by Walker is from a letter Einstein wrote to Marić dated by the *CPAE* editors in the second half of May 1901. He was writing from Winterthur, Switzerland, where he held a temporary teaching position. At that time Marić, in Zurich, was preparing to take the Polytechnic's final diploma examinations for the second time, in July. Here is the full context of the quotation:

> How is your work going, dear sweetheart? Everything proceeding jolly well? Does old [Heinrich Friedrich] Weber behave decently or does he again have 'critical theorems [*Sätze*, better: *comments*]'? The local *Prof. [Gustav] Weber is very nice to me and shows interest in my investigations. I gave him our paper.* [note 5] If only we would soon have the good fortune to continue pursuing this lovely path together. But destiny seems to bear some grudge against the two of us. But this will make things all the more beautiful later on, when all obstacles and worries have been overcome. (*CPAE* 1, Engl., doc. 107)

The editors' note 5 to this letter in the German volume 1 indicates that the paper in question was that on capillarity, published on March 1, 1901. It was his only published paper at the time; as we have seen, Marić assigned it unequivocally to Einstein, telling Kaufler Savić how proud she was of him (M-KS, 70). Einstein still expresses the hope that they would have "the good fortune to continue pursuing this lovely path together." But this hope was not to be fulfilled.

I turn now to the most frequently cited among the thirteen instances listed by Walker, but one he did *not* quote in his letter to *Physics Today* in 1991:

> "*How happy and proud I will be when the two of us together will have brought our work on the relative motion to a victorious conclusion!*" (doc. 94)

This sentence appears in a letter to Marić dated by the Einstein editors as March 27, 1901 (*CPAE* 1, doc. 94). Numerous authors have concluded from this sentence that Marić was collaborating with Einstein on the development of the special theory of relativity (e.g., T-P 1990, 426; Walker 1990, 1; Zackheim 2000, 20; Chiu 2008, 39; Frize 2009, 285; Milentijević 2015, 72–73). Evidently none of them have examined the letters closely enough to be cognizant of the fact that when in August 1899 Einstein enthusiastically reported to Marić his ideas on the ether and on the electrodynamics of moving bodies, in her direct reply to his letter she showed no interest in them, and wrote only about personal matters (E-M, 10–11, 12–13).

For centuries, natural philosophers and scientists have formulated notions of relative motion. In more recent times, many writers, including Carl Neumann, Henri Poincaré, and Hendrik Lorentz, explored theoretical discussions about relative motion. But the frequent phrase "relative motion" in published works was used in Einstein's time in reference to motion relative to the hypothetical ether, rather than the motions of objects relative to each other as in the later relativity theory (Martínez 2009). Similarly, the young Einstein was already formulating ideas on relative motion with respect to the ether as early as 1895, when he was sixteen years old, in an essay titled "On the Investigation of the State of the Ether in a Magnetic Field" (*CPAE* 1, doc. 5). Shortly after this he pondered about what an observer might see when running alongside a light beam at the speed of light (Fölsing 1997, 46). His outside reading led him in 1899 to question the existence of the ether, the medium in which a light beam traveled, in a letter to Marić from his alpine vacation on August 10 (*CPAE* 1, doc. 52; E-M, 10). Then a month later, on September 10, he wrote her, "In Aarau I had a good idea for investigating the way in which a body's relative motion with respect to the luminiferous ether affects the velocity of the propagation of light in transparent bodies" (E-M, 14). This became his proposal to Professor Weber for his diploma thesis research. Weber rejected it, apparently because it was too close to an experiment already performed by Armand Hippolyte Fizeau in 1851.

By the time Einstein wrote the sentence quoted by Walker (above), he was continuing to grapple with motion relative to the ether. It is highly improbable that Marić could have made any contribution without an immersion in the subject akin to Einstein's, evidence for which is nonexistent. Given this, and the context in the letter in question, the notion suggested by Highfield and Carter that Einstein was endeavoring to draw Marić into his extracurricular researches during a period when she was emotionally distressed[1] is made eminently plausible by his reference on April 30, 1901, to her apparently being jealous of a close friend who had recently married (E-M, 47; see also M-KS, 64–65). This was at a time when the exigencies of their respective situations meant that she was separated from Einstein for a lengthy period and longing to be with him, with a wedding date still undetermined. Soon after writing the sentence in question, Einstein stayed with his parents for several weeks (late March to early May 1901) and then held temporary teaching positions in Switzerland from May to December 1901, with a short break in July to holiday with his parents (E-M, 34–46, 51–72). During the latter part of this period he was again working on relative motion, as is evident from a letter he wrote to Marić from Schaffhausen dated December 17, 1901:

> It's really a screamingly funny life that I am living here, completely in Schopenhauer's sense [of solitude]. ... I am now working very eagerly on the electrodynamics of moving bodies, which promises to become a capital paper. I wrote to you that I doubted the correctness of the ideas about relative motion. But my doubts were based

solely on a simple mathematical error. Now I believe it more than ever. (*CPAE* 1, Engl., doc. 128)

Despite Einstein's optimism for the paper, it still had a long way to go. Over the next four years he struggled with a number of complicated ideas about relative motion and the ether before finally arriving in 1905, partly as a result of intense discussions with Michele Besso, at the foundation of what became the special theory of relativity (Fölsing 1997, 176–177; Martínez 2009, 240–280). Regarding the sentence quoted by Walker, Martínez writes:

> Non-specialists quickly concluded that this letter refers to the theory of relativity. ... But wait. The letter was written in 1901, and Einstein had no concept of the theory that he later formulated which became known as relativity. At that time he still believed in the invisible ether and sought ways to detect its relative motion experimentally. ... Only in spring 1905 did he abruptly formulate a radically new theory that became known as special relativity, after ten years of reflection, including more than seven years of intensive struggles. (Martínez 2011, 194–195).

OTHER INSTANCES LISTED BY WALKER

As in the above cases, the remaining six Einstein quotations listed by Walker in his footnote suggest scientific collaboration between Marić and Einstein, but other than reading advanced textbooks together, the existence of any actual research collaboration is not indicated. In document 50 (*CPAE* 1), a letter from Einstein to Marić in August 1899, Einstein mentions that he is studying Helmholtz's writings on "atmospheric movements." A little later he expresses regret that she is not with him while he is doing so. He adds that he finds the "working together" (*Zusammenarbeiten*) with her wholesome and less dry than reading the books alone. This shows that they did read and discuss this book together on an extracurricular subject, but it says nothing about her contributing to his theoretical ideas.

In his next two letters following document 50, Einstein alludes to material in Helmholtz's writings, and then in document 57 (cited by Walker)—Einstein to Marić, September 28?, 1899—Einstein writes that when they are together again in Zurich "we will start immediately with Helmholtz's electromagnetic theory of light." Again he is alluding to material they will read together, not necessarily researching together.

The remaining letters cited by Walker contain more of the same. Document 93 is Einstein's letter to Marić on March 23, 1901, written from Einstein's home in Milan. It contains a long paragraph in which Einstein recounts "an interesting idea" he has had on the latent energy of heat in solids and liquids in which he conceives these as the kinetic energy of electrical resonators. At the end of the paragraph he muses about the specific heat of glass and finishes with the suggestion that Marić (now back in Zurich) find some literature on this. (Since he was staying with his parents in Milan for some time during the spring of that year [E-M, 36, 38, 42, 44, 46] it was convenient for him to ask Marić to seek out the required information in Zurich, quite possibly in the Polytechnic library.)

Shortly afterward, in the letter dated March 27, 1901 (doc. 94, discussed above), he includes a paragraph describing his latest idea on "the problem of specific heat." Again, on April 10, 1901 (doc. 97), he provides another report on his latest notions "because I see that you like that," after mentioning earlier that he had second thoughts about "my idea about the nature of latent heat in solids" (docs. 94, 96).

In document 102, a letter to Marić on April 30, 1901, also from Milan, Einstein reports that he is again studying Ludwig

Boltzmann's theory of gases but thinks "however that in O. E. Mayer there is enough empirical material for *our investigation*." He adds that if Marić goes to the library she can check this, but that it can wait until he is back in Zurich. As noted by the editors of *CPAE* 1, these references are to ideas he was developing for his first doctoral dissertation (see *CPAE* 1, ed. note, 265–266). As we have seen in a letter (cited above) to Kaufler Savić in December 1901, Marić emphatically assigned the dissertation work to Einstein, indicating that if she gave any assistance it was marginal. It is noteworthy that in this instance we have Marić's letter in direct reply to Einstein's, in which he devotes a paragraph to an idea he had relating the emission of light to the kinetic energy of molecules, linking this to "our conservative molecular forces." In her reply, however, Marić provides no response to Einstein's discussion of physics, focusing only on an upcoming meeting with him. The only mention of Einstein's previous letters occurs when she tells him on May 3, 1901, how happy she always is with his letters, "which are full of ardent love, and which show her [Marić] that you are her dear sweetheart from before" (*CPAE* 1, doc. 105).

The letter to Marić on May 28?, 1901, document 111, opens with Einstein's enthusing about Philipp Lenard's 1900 paper on the photoelectric effect, followed by happiness at the news of her pregnancy and reassurances of his continuing love and support for her. He then adds: "Imagine how lovely it will be when we will again be able to work together totally undisturbed, and nobody will any longer be able to interfere! You will be amply compensated for your present worries by a lot of joy, and the days will peacefully pass by, undisturbed and unhurriedly" (*CPAE* 1, Engl., doc. 111).

Einstein is recalling their days as students, when they studied and worked together, or, as Einstein expressed it to Marić in

December 1901, when she had been his "student" (*CPAE* 1, doc. 130). At the time of this letter she was still working on research on heat conduction that she intended to include in a doctoral dissertation and had not yet taken the diploma exams for the second time. Einstein still held hopes of a future life together devoted to science.

SUMMING UP

We have not seen in Einstein's early letters to Marić—despite his occasionally writing "we" or "our" in the context of ideas or papers—any evidence that Marić made appreciable contributions to the researches he conducted leading to his first published paper, to his doctoral dissertation, or indeed to his ongoing research in other areas. Einstein used these first-person plural pronouns on eight occasions in relation to his extracurricular researches (including for his current doctoral dissertation), and these, with one exception, were during a short period between March 27 and December 12, 1901 (E-M, 36, 39, 41 [twice], 45, 47, 52, 68). The first of these was in connection with results relating to capillarity that *he* had obtained. The second was the isolated sentence we have examined above where he wrote of "our work on relative motion" with no mention of specific ideas, whereas he used first-person singular pronouns (*I* and *my*) in discussions of this same topic on seven occasions (E-M, 10–11, 14, 15, 46, 69, 71, 72). The third was in a context where all the ideas he expounded were evidently his (on his theory of molecular forces applied to gases), as was manifestly the case on two of the other occasions that he wrote on this subject (E-M, 32, 68). The fourth was in connection with his capillarity paper, and the fifth in the context of his own ideas on molecular forces. The sixth was preceded by his writing, "The local Prof. Weber is very nice to me and is interested in *my* investigations" (emphasis added) in regard to molecular forces. The seventh again related to his capillarity paper, and the last one was in the context of his discussing the consequences of a fresh idea that *he* had come up with regarding molecular forces. On all other occasions when he used first-person plural pronouns (*we* and *our*) it was in the context of their respective doctoral dissertations, with no indication that they had collaborated on these (E-M, 30, 32, 63; see also M-KS, 69).

Never cited by proponents of the claim that the early Einstein-Maric correspondence demonstrates that there was close scientific collaboration between the couple is the fact that Einstein used first-person singular pronouns (*I* and *my*) in relation to his exploration of extracurricular physics on some twenty occasions (E-M, 8. 10–11, 14, 15, 22, 32, 37, 40, 41, 43, 45, 47, 53, 55, 56, 59, 68, 69, 71, 72, 75–76). Typifying the confirmation bias of such proponents, and notwithstanding the absence of substantive evidence that Marić contributed to Einstein's investigations into advanced theoretical physics during this period, Milentijević (2015, 125-126) contends that "Albert's letters are replete with references to their close work together on research projects generated by his new ideas [which] reflect the fact that Albert relied on Mileva's close collaboration on a score [*sic*] of scientific endeavors …"

In conclusion, it is important to reiterate that Einstein's use of inclusive language was almost exclusively during a period when he was still maintaining his hopes for a life working together with Marić on physics, or as he expressed it on December 28, 1901: "When you're my dear little wife we'll work on science together so we don't become old philistines, right?" (E-M, 72–73; see also 52). Nevertheless, I am not asserting that the above analysis of Einstein's letters entirely precludes the possibility that Marić made verbal contributions of her own during the times when he discussed his extracurricular ideas with her directly; he was always bursting to communicate about these ideas with someone, such as his classmate Marcel Grossmann (*CPAE* 1, docs. 100, 122). But how much she may have contributed, or how little, we can never know unless fresh documents are found that throw additional light on this subject.

Of the six foundations listed by Trbuhović-Gjurić to support her conclusion that Marić's contribution to Einstein's creative work was large and significant, the fourth is her claim that Marić collaborated with Einstein during the period when they were studying together and during their marriage. I examined the first part of this contention in chapter 7 by scrutinizing the claims of collaboration made in relation to the early Einstein-Marić letters that were first published in 1987. I now turn to the second part of Trbuhović-Gjurić's contention, collaboration during their marriage.

Einstein and Marić were married in Bern, Switzerland, on January 6, 1903. On July 29, 1914, soon after the family moved to Berlin, they separated, and Marić and their two boys returned to Zurich for good. They were divorced in 1919. Their more than eleven years living together encompassed Einstein's "miracle year" of 1905, during which he completed his second and successful dissertation, along with his famous papers on atomic theory (Brownian motion), quantum theory, the special theory of relativity, and his famous equation $E = mc^2$. The period also encompassed a considerable proportion of the many years during which he was extending special relativity to what became his greatest achievement: the general theory of relativity, which included gravitation, published in 1915. I begin with the "Joffe

Story" involving the alleged co-authorship of Einstein's famous papers of 1905.

THE JOFFE STORY

One of the most frequently cited arguments that Mileva Einstein-Marić collaborated with Einstein in the creation of his theories during their marriage is the claim that the eminent Soviet physicist Abram Joffe saw the original manuscripts of Einstein's three most famous papers of 1905, and that the names of both Einstein and Marić were on them as co-authors.

The three papers, all published in a single volume (vol. 17) of the German journal *Annalen der Physik* (Annals of Physics), concerned his work on the quantum theory (including the photoelectric effect), Brownian motion (atoms), and the special theory of relativity. All appeared with Albert Einstein as the sole author. They may be found in facsimile reproduction in volume 2 of the *Collected Papers of Albert Einstein* (*CPAE 2*, docs. 14, 16, 23), and in the accompanying English translation volume. In the next volume (18) of the *Annalen* that same year, Einstein appeared as the sole author of the paper in which he presented his equation $E = mc^2$, though with different symbols (*CPAE 2*, doc. 24). This was an offshoot of the relativity paper. He also submitted his successful doctoral dissertation to Zurich University that year, which was published as a monograph (*CPAE 2*, doc. 15) and in the *Annalen* with slight revisions (not in *CPAE*), both in 1906.

The Joffe Story, regarding the Einstein-Marić co-authorship of the three 1905 papers (sometimes extended to all five publications), originated, as did many other questionable contentions, from Trbuhović-Gjurić's biography of Mileva Einstein-Marić. In it she writes (in translation):

The distinguished Russian physicist, director of the Physical-Technical Institute, later the Institute for Semiconductors at the Academy of Sciences of the USSR in Leningrad, Abraham [Abram] F. Joffe (1880–1960), pointed out in his "Recollections of Albert Einstein" that Einstein's three epoch-making articles in Volume 17 of "Annalen der Physik" of 1905 were signed in the original "Einstein-Marić." Joffe had seen the originals as an assistant to Röntgen, who belonged to the Board of the "Annalen," which had to referee contributions submitted to the editors. For this work Röntgen engaged his summa cum laude student Joffe, who in this way gained sight of the manuscripts, which are no longer available today. (T-G 1983, 79; 1988, 97)

It is immediately obvious that Trbuhović-Gjurić does not quote Joffe's actual words from his article--she quotes only the title (in translation). Nor does she distinguish between her own speculative contribution and information purportedly coming from Joffe. The same holds for Evan Harris Walker (1991, 123) who wrote that he had actually seen the original passage in the Russian journal. Not until 2002 did an English translation finally appear in print, albeit in an anti-Einstein diatribe (Bjerknes 2002, 196), later reprinted in Stachel (2005, lvi). Further translations appeared in Krstić (2004, 113–114), and Milentijević (2015, 123). Given the extreme rarity of public knowledge of Bjerknes's self-published book, or of Krstić's 2004 biographical volume that contains his own English translation of the relevant passage in Joffe's 1955 article (Krstić 2004, 113–114), prior to 2005 authors choosing to write about this subject had to rely solely on the problematic account provided by Trbuhović-Gjurić, or more frequently, her story as paraphrased for the benefit of English-speaking readers by Senta Troemel-Ploetz (1990, 419).

So the first thing that should be done in examining Trbuhović-Gjurić's claims is to see what Joffe actually wrote in his article.[1] It was written as a memorial to Einstein following his death in 1955. As translated from the original Russian:

> For physics and especially for the physics of my generation—that of Einstein's contemporaries, Einstein's entrance into the arena of science is unforgettable. In 1905, three articles appeared in the "Annalen der Physik," which began three very important branches of 20th century physics. Those were the theory of Brownian motion, the photon theory of light, and the theory of relativity. The author of these articles—an unknown person at that time, was a bureaucrat at the Patent Office in Bern, Einstein-Marity (Marity—the maiden name of his wife, which by Swiss custom is added to the husband's family name). (Joffe 1955, quoted in Stachel 2005, lvi)

It is apparent from his statement that Joffe is celebrating Einstein, "an unknown person [singular] at the time" as "the author [singular] of these articles." He does not claim that two people named Einstein-Marity appeared on the original manuscripts, nor does he claim that he had seen the manuscripts of these papers or any name on them. He states merely that "a bureaucrat" in the patent office—"Einstein-Marity"—was "the author of these articles." In parentheses he further states that the hyphenated name adheres to a Swiss custom by which the husband adds the maiden name of his wife to his own surname. ("Marity" is the Romanized transliterated version of "Marić" that appears on her Swiss marriage certificate (T-G 1983, 64; 1988, 82), and it was this version, transliterated again into the Cyrillic "Марити," that Joffe used in his article [Stachel 2005, lxxi, n.117]; we will return below to the vexed issue of this hyphenated name.)

What of the rest of Trbuhović-Gjurić's assertions in the paragraph quoted above that are obviously *not* in Joffe's statement? Trbuhović-Gjurić writes that Joffe was a scientific assistant to Wilhelm Röntgen (the discoverer of X-rays), and that, as a member of the editorial Board of *Annalen der Physik*, he evaluated the three papers and showed them to Joffe. However, she provides no substantiation for her claim that Röntgen and Joffe were involved in reviewing the manuscripts. There are good reasons to believe that this is nothing but evidence-free speculation on her part.

The editor of *Annalen der Physik* at that time was Paul Drude, and his adviser on theoretical physics was Max Planck, both of whose areas of expertise were in theoretical physics. Drude was the author of books and papers on electromagnetic theory, to which Einstein's relativity paper was related; and Planck had published on blackbody radiation, a topic relating to Einstein's paper on quantum theory. Thus, there was no reason for them to have shown the 1905 manuscripts in question to Röntgen, who was an experimental physicist, nor does any other evidence indicate that he had sight of the manuscripts.

The authors of the foremost account of the development of theoretical physics in Germany wrote regarding the editorial practices of the *Annalen der Physik* in 1905:

> As the advisor on theoretical physics for the *Annalen der Physik*, in 1905 Planck was already familiar with Einstein's work. For five years, Einstein had regularly submitted papers to this journal, the most important of which treated thermodynamics and statistical physics, subjects of particular interest to Planck at that time. Einstein extended these studies to a related interest of Planck's, blackbody radiation, in 1905. Einstein's relativity theory of the same year set Planck to work; it was the subject, Max Born observed, that "caught Planck's imagination more than anything else." (Jungnickel and McCormmach 1986, 248; quoted in Stachel 2005, lix-lx)

To add to the unlikelihood that Drude sent the manuscripts to Röntgen, we are supported by Joffe's own account of his academic work at precisely the time in question. In his book

Begegnungen mit Physikern (Meetings with Physicists) (Joffe 1967), a German translation of the Russian original (Joffe 1962), he describes his experience as a graduate student with Röntgen. He reports that the latter suggested to him that when he defended his doctoral dissertation in May 1905 he should discuss what one would now describe as the prehistory of the special theory of relativity, specifically the Lorentz and FitzGerald contraction equations (Joffe 1967, 23). Significantly, he makes no mention that Röntgen showed him Einstein's 1905 relativity paper shortly afterwards. Had he had the opportunity to see the original manuscript at that time, it is inconceivable that he would not have mentioned such a momentous experience in this context. Later in the book Joffe acclaims Einstein for his creation of the theory of relativity and for his wider influence on the physical worldview (Joffe 1967, 92). He goes on to identify some of Einstein's other achievements, among which he includes his work on Brownian motion and what became the photon (light quantum) theory of radiation. This covered the three 1905 papers which proponents of the Mileva Story claim were co-authored by Marić, citing Joffe as the source for this contention.

It is unsurprising that virtually all the authors who have written about Marić in recent decades have no knowledge of the scholarly material alluded to above, and it is regrettable that they have taken on trust everything Trbuhović-Gjurić wrote about the Joffe story. Most often the story will have been encountered in the more convenient English language version provided by Troemel-Ploetz (1990, 419), which glosses over problematic aspects of the original account, such as Trbuhović-Gjurić's failure to quote Joffe's actual words, that one would hope the perceptive reader would recognize. One author, who had the advantage of being able to read Trbuhović-Gjurić's account in German, is Andrea Gabor. After writing that "the original version of Einstein's three most famous [1905] articles … were signed Einstein-Marity," she stated that "Abraham [Abram] F. Joffe, a member of the Soviet Academy of Sciences, claimed that he saw the original [1905] papers when he was assistant to Wilhelm Röntgen, who belonged to the editorial board of *Annalen der Physik*, which published the articles." She went on to add an inaccuracy of her own to muddy the waters even further: "An article in a 1955 Soviet journal of physics quotes Joffe, now deceased, as ascribing the 1905 papers to 'Einstein-Marity.'" (Gabor 1995, 20) Also published in 1995 was an error-strewn account of the story, derived this time from Troemel-Ploetz's misleading version, by the well-known British feminist sociologist Hilary Rose (1995, 143). In this way Trbuhović-Gjurić's gross distortion of Joffe's paragraph has become widely disseminated.

One final point. In an article with the title (translated from the German) "'The parents' or 'the father' of relativity theory" Margarete Maurer (1996, 24) adds an additional argument purportedly in support of Trbuhović-Gjurić's account. She quotes a paragraph from a popular-science book by Daniil Semenovich Danin, published in Moscow in 1962, that in its essentials is very similar to Joffe's 1955 paragraph (above), except that Danin writes that the three 1905 articles mentioned by Joffe were "*signed* Einstein-Marity (or Marić—which was his first wife's family name.)" (Danin 1962, 57, emphasis added; see Stachel 2005, lv, for a translation of Danin's whole paragraph.) Maurer states that the context indicates that Danin had had conversations with Joffe, and that he was therefore transmitting information he had obtained from Joffe. However, as Stachel (2005, lv–lvi) points out, the articles could not have been signed "Einstein-Marity" *and* "Einstein-Marić," so Danin "has no clue how they were signed." Stachel goes on to observe that the close similarity between Danin's wording and Joffe's indicates that he was merely repeating an account he had read elsewhere, and, as so often happens in such cases, had mistakenly introduced a word ("signed") that was not in the original article.

FROM SINGLE TO DUAL AUTHORS

As we have already seen in Joffe's parenthetical remark, he used the idiosyncratic name "Einstein-Marity" to refer to the author of the three 1905 papers. He took it to be a Swiss custom for a husband to add his wife's maiden name following his surname, and he evidently wished to honor Einstein in this memorial with his full name in Swiss usage. While this usage did occur occasionally (e.g., the Swiss scientists Friedrich Miescher-His, Johannes Friedrich Miescher-Rüsch, and Julius Wagner-Jauregg), Joffe was wrong in thinking that it was a general Swiss custom.

Evan Harris Walker, having been informed by Senta Troemel-Ploetz of Trbuhović-Gjurić's passage regarding the Joffe statement, went even further with her erroneous assertion that Joffe had seen the papers. In his talk presented to a session at the American Association for the Advancement of Science in 1990 (see the Introduction), Walker stated: "If Joffe remembered that form of her name, it would have had to be because he had seen something that Mileva had signed herself, something that she signed 'Einstein-Mariti'" (Walker 1990, 15, using the transliteration from the Russian "Mariti" instead of "Marity"). As he subsequently argued in a letter to the journal *Physics Today*:

> Joffe's use of "Einstein-Mariti" ("Марити") agrees with Mileva's adoption of the Hungarianized spelling of her Serbian name Marić, a fact that Joffe would only have known had he seen the original signed by her, since this usage of "Mariti" apparently is not to be found in any of the Einstein biographies. (Walker 1991, 123)

Walker closed his AAAS talk by saying: "This, taken with all the rest, is compelling evidence that Joffe did see the original 1905 papers, and that the name there was 'Einstein-Mariti'!" (Walker 1990, 15).

Of course, what is compelling evidence for Walker may not be for others.[2] First of all, Walker is wrong that no Einstein biographies before Joffe's article contain the usage Marity for Marić. The second edition of the well-known documentary biography of Einstein by Carl Seelig, published in Zurich a year before Joffe's memorial, gives her name as "Mileva Marić or Marity" (Seelig 1954, 29; 1956a, 24).

Second, in his letter to *Physics Today* in 1991 Walker concludes, on the basis of an evidence-free presumption of Marić's requisite knowledge of electrodynamics, that "[i]t would seem then that Mileva Marić deserved to be a co-author, and her name should have appeared on the original 1905 [relativity] paper. ... And in fact it did" (Walker 1991, 123; also, Walker 1990, 14). As supposed evidence that "it did," he quotes at length the problematic passage from Trbuhović-Gjurić's book. It is remarkable that Walker made this statement despite the fact that, as he writes in his published letter to *Physics Today* (using the transliteration *I* rather than *J*), "I have found the paper of Ioffe's mentioned above." He goes on: "There Ioffe states, referring to the 1905 papers, 'Their author was Einstein-Mariti'—to which Ioffe added, believing that this referred to Albert Einstein alone—'an unknown clerk in the patent bureau in Berne (Mariti was his wife's surname)'" (Walker 1991, 123). Rather than accepting what Joffe actually wrote, he chose to read into Joffe's statement an imagined failure to recognize that Einstein-Marity referred to two authors, not one. How that would be is not clear. Did he now mean that Einstein-Marić (or Mariti) refers to Einstein *and* Marić? What about their first names? Instead of adhering to Joffe's actual statement, Walker chose to follow Trbuhović-Gjurić's erroneous account and the unjustified inference that Einstein-Marity referred to both Einstein and Marić, apparently because it, rather than the evidence or lack thereof, supported his own preconceived conclusion.

The paper published by Troemel-Ploetz in 1990 became probably the most widely known source for the claim of Marić's co-authorship based on a misreading of Joffe's writing Einstein-Marić for the author (singular) of the 1905 papers. It too derives from Trbuhović-Gjurić's erroneous account of Joffe's original statement, but with some errors of her own that mislead her readers even more than Trbuhović-Gjurić's report. She writes the following about the Trbuhović-Gjurić passage, without quoting the passage itself:

> Much more disastrous and devastating, however, is what happened to the five articles that appeared in 1905 in the Leipzig *Annalen der Physik*. Two of them, including his 21-page dissertation, were written in Zurich. It's an open question how much Mileva Einstein-Marić contributed to them. ... The other three articles published in Vol. XVII of *Annalen der Physik* were written in Bern while Albert Einstein was at the Swiss Patent Office and were written together with his wife. He later received the Nobel Prize for "Einen die Erzeugung und Verwandlung des Lichtes betreffenden heuristischen Gesichtspunkt" [paper on the quantum hypothesis]. "Elektrodynamik bewegter Körper" contains the special theory of relativity. Abram F. Joffe, the famous Russian physicist who was then an assistant to Röntgen (a member of the editorial team that examined the articles sent to *Annalen der Physik* for publication) wrote in his *Erinnerungen an Albert Einstein* [ref: in original: Joffe 1960 (actually referring to Joffe 1955)] that the original manuscripts for these two and also for a third paper were signed Einstein-Marić [ref: T-G 1983, 97 (actually 79); T-G 1988, 97]. Would the male editors have

dropped the name of a male co-author, or that of a woman who was not the author's wife? Would not a male co-author have protested against his name being dropped in the publication and would he not have asked for some form of reparation? The manuscripts, together with all the notes for these three papers, are no longer extant. The *New York Times* of February 15, 1944, wrote about the manuscript of the theory of relativity that Albert Einstein "had destroyed the original after the theory had been published in 1905. An $11,500,000 reward was promised to the person who could bring the original manuscript to the Library of Congress [ref: T-G 1983, 72 (1988, 90)]." (T-P 1990, 419)

It is evident from her article that Troemel-Ploetz had not consulted the original Russian source and was relying solely on Trbuhović-Gjurić's embellished account with its assumption that "Einstein-Marić" referred to two authors, not one.

In the above passage Troemel-Ploetz also repeats Trbuhović-Gjurić's mistaken assertion elsewhere that two of the five papers (supposedly the doctoral thesis and the Brownian motion paper) were written in Zurich (T-G 1983, 71–72; 1988, 90). In fact, all five were submitted from Bern in 1905 (*CPAE* 2, docs. 14, 15, 16, 23, 24), though the dissertation was not published until 1906.

As for the supposed destruction of the 1905 manuscripts, which Troemel-Ploetz regards as suspicious, the editors of *CPAE*, volume 1, state:

Einstein made no systematic attempt to preserve his papers before about 1920. Prior to that time, he routinely discarded manuscripts of published articles, and very few have been preserved. Einstein saved few letters addressed to him, though, fortunately, many of his correspondents kept the letters they received. Only a handful of early notebooks, containing lecture and research notes, have survived. (*CPAE* 1, xxvii)

THE OLYMPIA ACADEMY

Soon after Einstein settled in Bern in February 1902, he became friends with two other young men, Conrad Habicht and Maurice Solovine. Einstein had earlier met Habicht, a former philosophy student and now a math student at the University of Bern, while tutoring in Schaffhausen. Solovine had responded to one of Einstein's newspaper advertisements in Bern offering tutoring in physics. Rather than physics tutoring, the three so enjoyed discussing intellectual topics together that they inaugurated what they called (probably self-mockingly) the "Akademie Olympia" (Olympia Academy hereafter). (Fölsing 1997, 98–99; Isaacson 2007, 79–84). Together the three friends discussed philosophical issues and read and discussed a wide range of philosophical and cultural works, often over dinner and then in a member's apartment late into the night.

While Einstein was settling down in Bern during 1902, in the latter half of that year Marić (presumably having left their daughter Lieserl with her parents) spent two short periods staying close to Einstein's place before returning for good to Bern in mid-December 2002 (Krstić 2004, 87–90). There she and Einstein finally married on January 6, 1903, but without any family members in attendance. Aside from the town clerk, only Einstein's Academy colleagues were present as witnesses. After the wedding, they joined the newlyweds for a celebratory dinner (Isaacson 2007, 85). After their marriage Marić attended meetings of the Olympia Academy, and, according to Trbuhović-Gjurić (1983, 63; 1988, 81, 83), she fully participated in their reading of books by eminent authors. She writes that their discussions (purportedly also attended by Michele Besso and his wife) lasted well into the night, and she adds the following details: "Mileva, together with Einstein, accorded especial importance to Hume's question whether or not the Ego has a substance aside from itself. The two of them often had long, deep going discussions, about this" (T-G 1983, 63; 1988, 81). Trbuhović-Gjurić's source for this imaginative scenario is evidently Carl Seelig's biography of Einstein. Two sentences before the above quotation, she cites the following authors as well as their books read at Academy gatherings: Mach, Mill, Hume, Poincaré, and Dedekind. All of them, with their respective book titles, are cited in Seelig (1956a, 57). Aside from other errors in her account, she could not possibly have known what took place in private between Einstein and Marić, and has clearly embellished her account with an invented topic of their discussions, evidently inspired by Seelig's statement, "Hume postulated that there was substance neither in the ego, nor in nature" (Seelig 1956a, 61).

Charles S. Chiu provided one of the most extensive (and imaginative) subsequent discussions of Marić and the Olympia Academy in his book published in 2009. He writes:

> Mostly sitting silent in the background, Mrs. Einstein nevertheless follows the discourse of their friends with fascination. But because of her shy nature, she is most reluctant to join the current discussions; only when their guests have left does she become talkative and explain to Albert her personal conclusions regarding the constitution of the Ego, ethics, or the existence of ether in the universe. But in time, the other members of the "Academia" begin to notice the quiet Mrs. Einstein. The mathematical precision of her thoughts and her tendency to examine all theories according to their practical application become apparent even in the most unimportant conversations with Mileva. They invite her to participate in the diverse investigations of the "Academia." (Chiu 2008, 40)

Chiu's account follows so closely that given by Trbuhović-Gjurić (he even has Besso attending the Academy meetings) that

there can be no doubt as to his source for this scenario. But he
further embellishes Trbuhović-Gjurić's already overblown story
that put Marić at the very heart of the Academy's events by having her explaining to Einstein her conclusions regarding both
the ego and the ether in private conversations (see T-G 1983,
53; 1988, 58).

However, the Academy member Maurice Solovine, in his
later recollection of their meetings, made no mention of Marić's
mathematical discussions or of her having created the impression Chiu describes in the above quotation. Solovine wrote
(although he gets the date of the doctorate wrong): "When Einstein received his doctorate he married Mileva Marić, a young
Serbian woman with whom he became acquainted at the Polytechnic where she was a student. This event caused no change in
our meetings. Mileva, intelligent and reserved, listened attentively to us, but never intervened in our discussions" (Solovine
1956, xii).

THE LITTLE MACHINE

Although Einstein is well known as a theoretical physicist, he
also displayed an interest in electrical technology that may have
had its roots in his family background (Fölsing 1997, 9, 28–29,
35; Isaacson 2007, 11–12, 23, 143). One of the two top course
grades of 6 that he received while studying at the Zurich Polytechnic was for Electrotechnical Laboratory (appendix C). (The
other was for Scientific Projects in Physics Laboratories.) After
leaving the Polytechnic he evaluated many technical inventions
during his years of work at the Bern patent office, and later
served as an expert consultant on patents (Fölsing 1997, 105).
Moreover, in a letter written in 1930 he stated that he had "never
ceased to concern [himself] with technical matters" (Fölsing

1997, 105), and he occasionally collaborated with experimentalists such as Wander J. de Haas (see, e.g., *CPAE* 6, doc. 14). Against this background it should be no surprise that on July 15, 1907, though deeply engaged on his theoretical research, he wrote to Conrad Habicht, his fellow member of the Olympia Academy, and Conrad's brother Paul, "I have discovered another method for the measurement of very small amounts of energy" (*CPAE* 5, doc. 48). He had invented a device that amplified a very small initial voltage through electrostatic induction into an output voltage that could then be measured using a common electrometer (*CPAE* 5, ed. note, 51). The device, what they called a "little machine" (*Maschinchen*), could be useful for studying extremely small voltage fluctuations as low as 0.0005 V (*CPAE* 1, doc. 39, p. 396). Paul Habicht, who had a workshop in Schaffhausen for making and selling scientific instruments, worked on building a prototype of the new invention for possible sale. The credit for this invention has become a source of controversy.

MARIĆ'S ROLE

Following her discussion of the Olympia Academy, Trbuhović-Gjurić turned immediately to Marić's purported role in the invention of the little machine:

> Together with Paul Habicht she [Marić] began to work on the construction of an electrostatic influence machine for the measurement of small electric voltages by means of multiplication. It took a long time, not only because she had so much to do, but also mainly because of the thoroughness with which she followed all possibilities for perfecting it. When both were satisfied they left it to Albert to describe this apparatus as a patent expert. (T-G 1983, 65; 1988, 83)

Aside from the usual lack of a source for this statement, Trbuhović-Gjurić's account finds no support in the documents relating to the construction of the device. Moreover, contrary to the above account, Carl Seelig (1954, 73; 1956a, 60) whose short "Documentary Biography" of Einstein is generally reliable, writes that Marić's contribution to the work on the little machine was marginal: "Their [Einstein's and the Habicht brothers'] attempts to perfect it with occasional help from Mileva lasted several years."

As usual, Troemel-Ploetz recycles Trbuhović-Gjurić's account of the development of the device without caveats in her 1990 article (T-P 1990, 418–419). But the original documents pertaining to this episode (published in *CPAE* 5 in 1993) tell a very different story. This published material includes sixteen letters exchanged between Einstein and the Habicht brothers during the years 1907 to 1911 in which the device is discussed.[3] The letters follow Einstein's presentation of the device in papers published in 1907 and 1908 (CPAE 2, docs. 39, 48), and provide a well-documented account of its development (see also Maas 2007; *CPAE* 5, ed. note, 51–55). In none of these letters and, in particular, the six letters from Paul Habicht (who was attempting to build the device) to Einstein giving details of the stages in the construction of the device, is there any mention of Marić playing a role.

In a letter from Einstein to his friend Jakob Laub in November 1908, which appeared in the same volume, Einstein wrote, "The Maschinchen is ready and works well for higher voltages. In order to test it for voltages under 1/10 volt, I built an electrometer and a voltage battery. You wouldn't be able to suppress a smile if you saw the magnificent thing I patched together myself" (*CPAE* 5, doc. 125). In a letter addressed to Conrad Habicht a few months later (March 1910), Einstein wrote: "I'm inviting you here to stay with us, so the two of us together can do the final experiments with the Maschinchen and piece together the paper. ... We have a spare room for you and your brother" (*CPAE* 5, doc. 198). Ad Maas (2007, 309), curator of the Leiden Museum Boerhaave, the Netherlands, who has researched the origins and development of the little machine concludes: "There is no evidence that Einstein's wife was involved in the development of the little machine as has been alleged by Trbuhović-Gjurić." The above documented information does not preclude occasional help from Mileva (Seelig), but it does contradict the notion that Marić built the device together with Paul Habicht.

Despite the publication of the relevant letters in 1993 (*CPAE* 5), several authors writing after that year repeated Troemel-Ploetz's repetition of Trbuhović-Gjurić's account. One was Hilary Rose, who wrote in a book chapter with a short section devoted to Marić that "[t]wo key episodes document the process by which her work, if not actively appropriated, was certainly lost by her to him [Einstein]." One of these, according to Rose, was that "two [*sic*] of the originally submitted [1905] manuscripts were signed also by Mileva, but by the time of their publication, her name had been removed." The other one was reported by Rose as follows:

> In one episode Mileva, through the collaboration with a mutual friend, Paul Habicht, constructed an innovatory device for measuring electric currents. ... Having built the device the two inventors left it to Einstein to describe and patent, as he was at that time working in the patent office. He alone signed the publication and patented the device under the name of Einstein-Habicht ... Later when the marriage had collapsed she found that the price of her selfless love ... was that her work had become his. (Rose 1994, 143)

None of this is accurate, most notably the assertion that the two inventors were Marić and Paul Habicht, with Einstein playing no role. Although Einstein published the papers in

1907 and 1908 outlining his method of measuring small quantities of electricity (*CPAE* 2, docs. 39 [end] and 48), the machine was not built to Einstein's satisfaction until 1910, by which time he was no longer working at the Bern patent office. Einstein helped draft a technical article describing the device, but when the prototype was completed in March 1910 he allowed the Habicht brothers to take credit for its design and testing. The article was published under the names of Conrad and Paul Habicht alone (Habicht and Habicht 1910). In the article they stated they had applied for a patent, but Maas writes that it is uncertain if one was actually granted (Maas 2007, 314–319; *CPAE* 5, ed. note, 53).

THE MISSING PATENT

Troemel-Ploetz, again following Trbuhović-Gjurić, writes that at the time Einstein wrote his 1908 paper on the little machine, "he had the apparatus patented under the name *Einstein-Habicht* [italics hers] (Patent No. 35693)" (T-P 1990, 418–419). Trbuhović-Gjurić actually wrote only that Einstein had *applied* for a patent (*zum Patentieren vorgelegt worden*), but she does cite the alleged patent number, as usual without any source reference to confirm its existence and help others locate it (T-G 1983, 65; 1988, 83). But in a letter to Conrad Habicht in December 1907 Einstein reported, "I have dropped the patent, mainly because of the manufacturers' lack of interest" (*CPAE* 5, doc. 69). This apparently accounts for the fact that all attempts to locate this patent have been unsuccessful, despite the reported patent number.[4]

MORE REPORTS OF COLLABORATION

In the literature on Mileva Marić there are several other accounts of reports that claim she collaborated with Einstein on his scientific research. These come from family members or acquaintances who gave (mostly hearsay) accounts of events that supposedly occurred during visits that the Einsteins paid to Marić's parents in 1905 (and allegedly in 1907); from the same people's purported recollections that allude to a narration Marić's brother Miloš gave to his father after returning from a short stay with the Einsteins in Bern in 1904; and from a student who lodged with the Einsteins when they resided in Zurich either in 1909–1910 or 1912–1913, whose comment pertaining to an aspect of the Einsteins' domestic life was passed on some fifty years later by the student's daughter. In this section I explore the provenance and reliability of this seemingly impressive list of reports.

VISITS WITH THE MARIĆ FAMILY

In one passage in her book, Trbuhović-Gjurić reports conversations that took place between Einstein and some student friends of Marić's brother Miloš when the Einsteins visited the Marić family home in Novi Sad supposedly in 1907 (see below). Many of the students, she writes (T-G 1983, 77; 1988, 95), knew of his papers "and considered them to be the result of the collaboration between Albert and Mileva," which hardly constitutes evidence. Likewise, Krstić reports that the couple, with their young son Hans Albert, visited Novi Sad in 1907. He writes that Marić's orphaned cousin Sofija Golubović (née Galić), who was living with the Marić family in Novi Sad at the time, told him in 1961 that "Mileva and Albert would debate, work on mathematical computations and write together" (Krstić, 2004, 127).

However, according to Michele Zackheim in her book recounting her efforts to trace the fate of the couple's daughter Lieserl, Einstein made no such visit in 1907: "By summer 1907,

Albert was so busy with his job and his many projects that Mileva went alone with Hans Albert to visit her parents" (Zackheim
2000, 63). A little later in her book Zackheim was more specific.
She writes that in August 1907 the Einsteins went on holiday in
the Alps near Bern and stayed at the village of Lenk. She then
states, without providing a source, that "within a week of returning home, Mileva left again with Hans Albert to visit her parents
in Novi Sad" (Zackheim 2000, 64).[5]

So who is right about Einstein's visit, or non-visit, to Novi
Sad in 1907? The now available evidence supports Zackheim's
statement that Marić went with Hans Albert alone in late summer 1907, while Einstein remained at home in Bern. A confirmation of the first of Zackheim's contentions is provided by a
letter Einstein wrote from Bern on July 15, 1907, in which he
told Conrad and Paul Habicht that he would be going to Lenk,
Switzerland, "on August 1 with wife and child for a ca. 10-day
vacation" (*CPAE* 5, Engl., doc. 48). On August 7 Einstein sent a
letter from Lenk to Wilhelm Wien in which he wrote that he
was currently on "two weeks of vacation" (*CPAE* 5, Engl., doc.
52). (Fölsing [1997, 227] confirms that at the patent office Einstein had two weeks annual leave.) On August 11 Einstein wrote
Wien again, this time from Aeschi, Switzerland (*CPAE* 5, doc
53). The above being the case, it is not possible for Einstein to
have accompanied his wife and Hans Albert on a second vacation visiting her parents in the middle of August 1907. And
indeed, a photograph reproduced in Krstić's book (Krstić 2004,
129) with the caption "Einstein/Marić family photo taken during the 1907 visit to Novi Sad" shows Marić with Hans Albert,
her sister Zorka, her brother Miloš, and her young cousin Sofija
Galić—but not Einstein. (Zackheim likewise captions the same
photo as taken "circa 1907, Novi Sad" [Zackheim 2000, photo
section].)[6]

Another piece of evidence comes in a letter Marić wrote to Kaufler Savić from her parents' family estate in Kać, dated by Zackheim as being from the summer of 1907 (Zackheim 2000, 63). The letter is written throughout in the first person singular (it opens with the words "I have been with my parents for a week now"), indicating that Einstein had not traveled with her and Hans Albert on that occasion (M-KS, 87). Popović dates the letter "(1906?)," but that can hardly be correct, as neither Trbuhović-Gjurić nor Krstić (nor indeed Zackheim) mention any such visit in 1906. Nor is it likely to have been 1908, as Marić and Hans Albert visited her parents without Einstein over Easter of that year, as we know from a letter he wrote her from Bern (*CPAE* 5, doc. 96), and there is no documentation to indicate a second visit. Nor can it be from 1909, in the June of which year Marić sent a letter to Kaufler Savic from Novi Sad in terms that preclude its being in the same year as the one that Popović dates from 1906 (see contents of the latter letter above): "Only today I received your dear letter, and I got it here, where I have been staying for just over two weeks" (M-KS, 96). Taken together, the above items demonstrate that it can be stated with confidence that Zackheim was right to assign the date of the letter in question to mid-August 1907, and that Einstein did not accompany Marić when she visited her parents in that year.

As a coda to this section, it is pertinent to note that Zackheim (2000, 93) reports that "in early summer, 1933 ... [Einstein] and Helene Savić may have seen each other in Novi Sad where Helene was visiting friends ... Serbian historians are unyielding in their conviction that he was indeed in Novi Sad at that time. Two well-known and highly esteemed citizens of Novi Sad, Dr. Aleksander Moč and Dr. Koster Hadži, maintained that they spoke with Albert that summer at the Queen Elizabeth Café. ... Dr. Moč, for one, had known Albert for many years. The three gentlemen reportedly sat at a table near the sidewalk, talking about the postwar situation in Germany." However, the whereabouts of Einstein in the spring and summer 1933 has been chronicled in reputable biographies and there is no record of his having visited Serbia at that time (Fölsing 1997, 666, 671–674, 676–677). Isaacson (2007, 418–425), in particular, closely researched Einstein's several commitments and itineraries during the months in question and found no evidence of any trip to Novi Sad in 1933. The reported conviction of (at least some) Serbian historians that Einstein did go to Novi Sad in early summer should be treated with the utmost skepticism.

Even more unlikely is the widely believed story in Albania that Einstein visited that country in the early 1930s to obtain an Albanian passport to enable him to enter the United States. That Einstein had retained his Swiss citizenship, and had a Swiss passport, evidently doesn't make a dent in the will to believe patriotically-inspired stories of this kind. (See https://prishtinainsight. com/urge-fake-history/) An Albanian-American newspaper even quoted eyewitness testimony to "prove" that the story was true (http://gazetadielli.com/albert-einstein-in-albania-and-the -albanian-passport/).

THE 1905 VISIT

Since it is evident from the above that Einstein did not accompany Marić on her visit to Novi Sad in 1907 as Trbuhović-Gjurić and Krstić claimed (T-G 1983, 76–77; 1988, 95; Krstić 2004, 126–128), it follows that the events involving Einstein and Marić reportedly occurring in 1907 are either inaccurate or should be attributed to the securely documented visit in 1905.

The official chronology of Einstein's life (*CPAE* 11, p. 184) has Einstein's first visit to the home of his wife's parents in "late summer" 1905. He did not visit again until 1913 (*CPAE* 11, p. 195). According to Zackheim (2000, 61), in late August 1905 the couple spent the first week of Einstein's annual two-week holiday with Marić's friend Helene Kaufler Savić and her family, mostly at Kijevo, a lakeside village near Belgrade. For the second week they journeyed to visit Marić's parents in Novi Sad and also stayed at the Marić family's estate at Kać for several days. Similarly, both Trbuhović-Gjurić 1983, 74; 1988, 92–93) and Krstić (2004, 114–115) chronicle these two separate excursions in late summer 1905, embarked upon several weeks after Einstein had sent his relativity paper to the *Annalen der Phyik* at the end of June that year (*CPAE* 2, Engl., 171).

A hearsay account of supposed scientific collaboration between Einstein and Marić during their several days stay at Kać during their week's visit to Marić's parents in 1905 appeared in a film titled *One Stone—Einstein* (with English subtitles) released by a Serbian television company in 2006. In the film, Dragiša Marić, described as "a cousin of Mileva Marić," alludes to "grandparents Žarko and Rada [Marić]" and other unnamed relatives who reportedly witnessed Marić and Einstein working together at the family estate in Kać:

> As for ourselves, a part of our family, grandparents Žarko and Rada, and all those who lived on our salas [estate], witnessed the couple, who were scribbling something all day long, working. Day and night they were together, and doing well. They spent summers here, and as their biographers tend to say, "they spent their happiest and most inspiring summers here." Grandpa Žarko later told us that they were working late into the night. All the theories born in that miracle year of 1905 were born here, in Kać, to be taken abroad and published. There are other proofs as well, but this is what we know by word of mouth. (Čvorić 2006)

Figure 8.1
Mileva, Hans Albert, and Albert, 1904–1905.
Courtesy of ETH-Bibliothek Zürich, Bildarchiv (photo archive),
photographer unknown, Hs_1457–72. Public Domain Mark.

Krstić (2004, 116–117) similarly reports what he describes as "eyewitness descriptions of the couple's joint scientific work" during their short stay in Kać in 1905 that he obtained more than five decades later. However, he provides no specific information, writing only that he "spoke in detail with Mileva's close relative Žarko Marić (1880–1965) and his eldest daughter Djurdjinka (1903–1990), who were living in the villa *Kula* during all three [actually two] stays of the Einsteins in Kać." Though he does not state explicitly that it was Žarko who provided the eyewitness report, the reader is left to draw this conclusion. (Clearly it could not have come from his daughter Djurdjinka, since she would have been only about two years old at the time [Krstić 2004, 116].) He makes no mention of any other person to whom he "spoke in detail" about the Einsteins' stay at Kać, but he does claim (2004, 117, 120) that Marić "sewed for Hans Albert, calculated and wrote, and worked together with Albert, or talked and walked with him." A little later he writes that the couple "would sit in the garden and discuss some physics problems, which were not at all understandable to the others." However, he does not cite any specific individuals as the source of this information.

Interestingly, in Trbuhović-Gjurić's account (1983, 74–75; 1988, 92–93) she does not mention any reports of the Einsteins' working together on science during this stay in Kać in 1905, although she presumably had contact with the same people cited in the 2006 Serbian film and by Krstić during her extensive research that she undertook during the 1960s which included communications, both oral and written, with members and acquaintances of the Marić family (T-G 1983 and 1988, 6).

What are we to make of these hearsay accounts obtained some fifty-five years later as reported by Krstić and (apparently) nearly a century later in the case of the Serbian film—accounts conspicuously missing in Trbuhović-Gjurić's report of the same short period in 1905? To start with, it is hardly credible that the couple, on their first visit for one week in 1905 during which they introduced Einstein and the infant Hans Albert to Marić's parents, would have acted so discourteously, "scribbling something all day long, working ... day and night," as Dragiša Marić asserts. The statement that "all the theories born in that miracle year of 1905 were born here in Kać, to be taken abroad and published," if we are to treat it with a seriousness it manifestly does not merit, is contradicted by the dates of Einstein's papers. Of the five main papers of 1905, the ones on quantum theory and Brownian motion were already published by August. Einstein had completed his second doctoral dissertation in April 1905 but did not submit it officially to the University of Zurich until July 20, where the faculty unanimously accepted it on July 27—before the trip to Novi Sad (*CPAE* 2, ed. note, p. 170). Likewise, the paper on relativity theory was received by *Annalen der Physik* on June 30, well over a month before the visit with the Marić family. Einstein's brief, three-page paper on $E=mc^2$ was received by the *Annalen* on September 27, well after the Einsteins had returned to Bern.

Dragiša Marić's assertion that "this we know by word of mouth" amply serves to illustrate the observation of Charles Fernyhough (2012, 130–131): "Findings on rich false memories show that the misinformation is particularly strong when other people, especially family members, are providing the interjected information. Some benefits accrue to collaborative remembering." As for Krstić's much more limited claim, aside from the reservations that apply to hearsay reports relating to supposed events obtained many decades later, it is so vaguely (and tendentiously) worded that it scarcely counts as evidence.

There remain the accounts by Trbuhović-Gjurić (1983, 77; 1988, 95) and Krstić (2004, 126–128) of events that occurred during the (for Einstein) non-existent summer 1907 visit to Novi Sad. Trbuhović-Gjurić reports, as does Krstić, that Einstein several times visited the café "Königen [Queen] Elisabeth" and talked with friends of Marić's brother Miloš. The consistency of these two reports suggests they are authentic, but Krstić provides no source for his account, and Trbuhović only one, Dr. Ljubomir-Bota Dumić, a one-time student friend of Miloš's (see chapter 6). It would appear that whoever it was who provided the reports of the café conversations conflated the year in which they occurred with that of the Einsteins' 1905 visit. Neither author mentions alleged scientific collaboration between Einstein and Marić being reported during these conversations.

In regard to the supposed 1907 visit, Trbuhović-Gjurić again makes no mention of any family members having seen the couple working together on physics, but, as already noted, Krstić (2004, 127) reports that in 1961 Marić's cousin Sofija Golubović "told [him] that the Einstein couple and [Marić's brother] Miloš, now a soldier, used to talk about scientific themes, and that Mileva and Albert would debate, work on mathematical computations and write together." Given Krstić's above report of the Einsteins' stay for several days in 1905 in Kać, and this one relating to their very short stay in Novi Sad that could only also have been in 1905, we are being asked to believe that the Einsteins spent most of their week's holiday at Marić's parents' two homes working together on physics for many hours each day. This is made all the more unlikely by the fact that for several months earlier that spring and summer Einstein had labored on his time-consuming production of four major papers, and would hardly have wanted to spend another week of his well-earned vacation engrossed in yet more physics problems. Given all this, the recollection Krstić obtained from Golubović in 1961, as reported in his 2004 book, must be considered highly suspect. That Trbuhović-Gjurić makes no mention of having received such reports during her extensive research in the 1960s only serves to increase the doubts that the collaboration scenarios reported by Krstić are authentic.

In the context of the 1905 visit Trbuhović-Gjurić (1983, 75; 1988, 93) writes that Marić told her father, as witnessed, among others, by Desana Tapaverica, a relative and the wife of the mayor of Novi Sad Dr. Bala: "A short while ago we finished a very important work that will make my husband world famous." Trbuhović-Gjurić adds that "[m]uch later, she [Mrs. Bala] recalled these words and narrated them," though to whom she does not say. Krstić (1991, 93–94) reports the same story as told to her father "and [to] her good friend Desana Tapavica [*sic*] who was married to Dr. Bala, the mayor of Novi Sad." Krstić cites the sources of the report as Sida [Sidonija] Gajin in 1955 (whom he describes [2004, 5] as a godparent or sponsor of Marić's), and also Žarko Marić in 1961. In his 2004 book, Krstić (2004, 115) quotes the same sentence, with the addition of a second sentence: "He received his doctorate, too." This time Krstić (2004, 115, n. 252) appends a footnote in which he cites Sofija Golubović as an eyewitness to the incident, adding that he heard the same comment but without the second sentence from Sidonija Gajin and Žarko Marić, "who were told that by Mileva's father Miloš."

According to the report, Marić referred to work that had been completed very recently, which indicates that it was the special relativity paper, written during five or six weeks from late May to the end of June 1905 (Stachel 2005, 113). However, of the four papers that Einstein (*CPAE*, 5, doc. 27) described to Conrad Habicht in late May 2005, it was the one on the photon theory that he described as "very revolutionary," not the relativity paper, though at that time it was already in "rough draft." One wonders how Marić (or indeed Einstein) could predict at that stage that the latter paper would make Einstein world famous. (In fact, it was not until 1919 that he became world famous, but for the later much more ambitious general theory of relativity following its experimental confirmation.)

That, by Krstic's accounts, he obtained recollections by four people (witnesses or hearsay) of the sentence in question may indicate that Marić did tell her father the words as reported (or something like them), or it could be an example of what Fernyhough described as "collaborative remembering" by family members and friends, especially the reporting of the word "we." If the former (or possibly a mixture of the two), we can't be sure of the exact wording as it was recounted some fifty or more years later. From an historical point of view nothing definite can be concluded from this particular incident as described by Trbuhović-Gjurić and Krstić.

CHALLENGING THE ETHER

In her 1990 paper, Senta Troemel-Ploetz poses the question: "Why did he [Einstein] not acknowledge in public that it was she [Marić] who came up with the idea to investigate ether and its importance?" (T-P 1990, 418). For the contention implicit in this question she cites, as usual, Trbuhović-Gjurić's biography, where the latter quotes verbatim a hearsay statement that Einstein reportedly said to Marić's younger brother Miloš during the 1905 visit with the Marić family: "She was the first to direct my attention to the significance of the ether that was believed to pervade space" (T-G, 1983, 69; 1988, 87).

Miloš, later a medic in the Serbian army during World War I, was taken prisoner by the Russians in 1917. He remained in Soviet Russia after he was released at the end of the war and never returned home (M-KS, 119; Krstić 2004, 217–220), so it not possible that the above report about the ether came from him directly. As usual, Trbuhović-Gjurić provides no source for the alleged statement. Presumably it was handed down to relatives and friends of the Marić family whom Trbuhović-Gjurić met while researching her book. If so, its unreliability after so long a time should hardly need emphasizing, yet Trbuhović-Gjurić (and Troemel-Ploetz) accepted it, as other similar statements, without caveat.

In this instance there is not only no evidence to support the assertion that Marić directed Einstein's attention to the significance of the ether (a topic he could hardly have failed to have come across in his early readings of physics texts), there is conclusive evidence to the contrary. His interest in motion relative to the hypothetical ether is abundantly evident in several of the letters he wrote to Marić in their student days. For instance in September 1899 he wrote: "In Aarau [where he had recently visited] I had a good idea for investigating the way in which a body's relative motion with respect to the luminiferous ether affects the velocity of the propagation of light in transparent bodies" (E-M, 14). He had written even earlier a precocious essay with the title "On the Investigation of the State of the Ether in a Magnetic Field" in 1895, when he was only sixteen, more than a year before he had even met Marić (*CPAE* 1, doc. 5). In addition, as already noted in chapter 7, when Einstein discussed the same topic in a remarkably prescient passage in a letter written in August 1899, Marić failed to show any interest in it in her lengthy direct reply to his, writing only of personal matters (E-M, 10–11, 12).

THE VARIČAK REPORT

Moving forward to the end of first decade of the twentieth century, Trbuhović-Gjurić (1983, 87; 1988, 105) recounts another hearsay story, this time purporting to demonstrate Marić's mathematical participation in Einstein's work. This one concerns Svetozar Varičak, whom she reports as having lodged with the Einsteins when he was a student in Zurich. Trbuhović-Gjurić writes that Svetozar's father Vladimir, a mathematics professor at Zagreb University, had met Einstein at a mathematics conference in Berlin in 1910, on which occasion the professor mentioned in a private conversation that his son was going to Switzerland to study chemistry and had no place to stay. Einstein responded: "My wife is a Serb and provides board and lodging for students. When the time comes, your son can stay with us. I will speak to my wife about it." Trbuhović-Gjurić then states that Svetozar Varičak's daughter "recalled having heard her father recount that from time to time Einstein helped his wife with the housework because it troubled him that when she had completed the household chores she was still trying to solve mathematical problems in his notes after midnight."

Trbuhović-Gjurić's account of how Svetozar Varičak came to have stayed with the Einsteins is suspect on three counts. First, there is no record of Einstein's going to Berlin before 1912. Second, it is hardly likely that Einstein would have taken time out from his teaching duties at the University of Zurich (the relevant period is from October 1909 to March 1911) and his time-consuming researches in theoretical physics to attend a *mathematics* conference. Third, she provides a three-sentence *verbatim* response by Einstein to Vladimir Varičak's comment in what she reports was a *private* conversation, which invites the question: How could anyone possibly know this kind of detail when recounting the story to Trbuhović-Gjurić? In fact, she does not tell her readers, in relation to what is now third-hand (at least) hearsay, from whom she came to know the story of Marić's supposedly working past midnight on Einstein's "notes." She doesn't say she heard it directly from Svetozar Varičak's daughter, and if she had, one would be left wondering how she made contact with the daughter since there is no evidence that the latter was an acquaintance of the Marić family.

Einstein traveled to Berlin for the first time in April 1912 (*CPAE* 5, doc. 384; Fölsing 1997, 330), where he met with eminent physicists about a possible position in Berlin. He had already been appointed professor at the ETH and would soon be moving back to Zurich from Prague. On May 14, 1913, news of Svetozar appears in a letter from Einstein to Vladimir Varičak, in which Einstein reported, "Your young lad is a very keen student and always in good spirits" (*CPAE* 10, supplementary doc., p. 21). This indicates that when he was a student Svetozar evidently stayed with the Einsteins for a period of time, in which case he very likely would have conveyed to his daughter some years later his impressions of the Einstein household, but how Trbuhović-Gjurić came to hear of them remains a mystery.

The alleged recollection of Svetozar's daughter refers to "notes." Other than his lectures for a single course at the University of Bern as an instructor (*Privatdozent*), the first time Einstein required extensive, carefully prepared lecture notes was when he was appointed professor of theoretical physics at the University of Zurich, commencing on October 25, 1909 (Fölsing 1997, 251). Having been told in early May 1909 that he had been appointed to the post, he evidently set to work on preparing his lecture notes not long after that, as Marić reported to Kaufler Savić on June 25: "My husband is now very busy preparing his lectures." It was in the winter semester 1909–10 that she complained to Kaufler Savić about Einstein's neglect of her on account of his "damned science," and at the end of

December 1912 she wrote that Einstein "is entirely working on his problems. ... I must confess with a bit of shame that we are unimportant to him" (M-KS, 97, 102, 108). It seems rather unlikely that, at some time during this period or just after, Marić would have "work[ed] past midnight trying to solve mathematical problems" in Einstein's notes while he did some housework. There is clearly a kernel of truth in the report in that, when he was a student, Svetozar Varičak evidently did lodge with the Einsteins in Zurich in 1913 (though whether he did so also in 1910 as Trbuhović-Gjurić's states is unknown). Even so, hearsay of unknown provenance purporting to recount what Svetozar said to his daughter many decades earlier does not qualify as historical evidence.

Versions of the Varičak story told by other authors exemplify what happens when accounts are repeated. According to Troemel-Ploetz:

> A mathematician of the University of Zagreb recalled that Albert Einstein every now and then helped his wife doing the household chores because he felt sorry that after her housework was done, she had to do his mathematical problems till way past midnight. (T-P 1990, 426)

Now Svetozar Varičak was a student of chemistry at the time (it was his father who was the Zagreb mathematician). More importantly, Troemel-Ploetz's account misleadingly reads as if it were a direct, first-hand report on the activities of Einstein and Marić, rather than a report made by Svetozar to his daughter that reached Trbuhović-Gjurić many decades later.

Andrea Gabor writes in the chapter on Marić in her book *Einstein's Wife*:

> Svetozar Varičak, a student who lived with the Einsteins for several months in about 1910, remembered how Marić, after a day of cleaning, cooking, and caring for the children, would then busy

herself with Einstein's mathematical calculations, often working late into the night. Varičak said he remembered feeling "so sorry for Mileva" that he sometimes helped her with the housework. (Gabor 1995, 20)

Gabor, too, presents this story as if it came directly from Svetozar Varičak. Moreover, in the original story it was Einstein, not Varičak, who felt sorry for Mileva, and it was Einstein, not Varičak, who helped with the household chores. Again, both reports allude to Einstein's mathematical "problems" or "calculations," implying these were related to his current researches, whereas Trbuhović-Gjurić's report refers more specifically to Einstein's "notes." In Krstić's account, for which he cites Trbuhović-Gjurić, he does refer to Einstein's notes, as does Milentijević citing Krstić. However, both authors present the story as if it were historical fact, with no mention of the obscure provenance and problematic nature of Trbuhović-Gjurić's hearsay source (Krstić 2004, 141–142; Milentijević 2015, 143–144).

INVENTED DIALOGUE

Troemel-Ploetz's propensity to accept Trbuhović-Gjurić's assertions without reservation is further exemplified when she writes that "during their common student years his [Einstein's] own view of Mileva Einstein-Marić was that she would make a better physicist than many men" (T-P 1990, 419). The original passage in Trbuhović-Gjurić's book from which this derives included a purported exchange between Marić and Einstein in which Marić exclaimed, "I believe that I would be just as good a physicist as my male colleagues," to which Einstein replied, "a better one, better than many others" (T-G 1983 and 1988, 41). Apparently Troemel-Ploetz did not ask herself how anyone could possibly know what was said in a private conversation at that time. This supposed exchange is one of several scenarios and verbatim conversations attributed to Einstein and Marić during their early student years at Zurich Polytechnic in this section of Trbuhović-Gjurić's biography (T-G 1983 and 1988, 40–41).

As usual, Trbuhović-Gjurić does not cite a source for the scenarios in question. It turns out that this and the other exchanges between Einstein and Marić were reproduced word for word (translating from English into Serbian and German) from a fictionalized biography of Einstein written for children and young readers by Aylesa Forsee.[7] Her book was manifestly not intended to be treated as a nonfiction work for adults, as is evident from the numerous other imaginative scenarios and episodes with dialogue invented by the author throughout the book in order to engage young readers—*not* to contribute to Einstein scholarship.[8] Forsee (1963, viii) writes in her acknowledgments that Hans Albert Einstein cooperated "in clearing up conflicts in biographical materials." But an examination of the items cited here suffices to show that this "clearing up" could not apply to these scenarios with invented dialogue, which are obviously intended to render the story of Einstein's life more accessible to children and young readers. A typical example is the following exchange set in 1905:

"You should send this to the *Annals of Physics*," Mileva declared, when she had glanced through the article Einstein had written about his research on photons that he had titled "The Quantum Law of the Emission and Absorption of Light."

"I'm not sure the editor would be interested," said Einstein, thinking of the bulky *Annals of Physics* full of technical papers and peppered with footnotes and references. The editor accepted the article, but Einstein's theory that light was made up of photons aroused arguments. It clashed with the long-held and not completely disproved theory that light was transmitted in waves.

"Doesn't it bother you when critics attack your ideas?" asked
Mileva.

"*Macht nichts*—it doesn't matter," said Einstein with a shrug.
"Anyone who chooses to be an innovator must be prepared to follow a lonely path." (Forsee, 1963, 18)

That Forsee's book cannot be remotely considered an historical work for adult readers (nor was it so intended) is further
indicated by a passage in the book in which Einstein becomes
increasingly attracted to "a young, black-haired Serbian woman"
during their first months at the Zurich Polytechnic (Forsee
1963, 10; T-G 1983 and 1988, 40). "Mileva Marić seemed
unresponsive at first; but after Albert learned a few words in
Serbian, she started mastering German so she could talk with
him" (Forsee 1963, 10). This is obviously pure fiction. Marić
had studied German for many years at school; she also attended
the Higher Girls' School in Zurich for two academic years and a
summer semester course in medicine at the Medical School of
Zurich University prior to enrolling at the Polytechnic (Krstić
2004, 33). Her letters confirm her fluency in German. Even the
limited number of such imagined scenes reproduced in
Trbuhović-Gjurić's biography should have sufficed for later
authors to recognize their fictional character and prevented their
utilization of such material by later authors (see chapter 9).

Trbuhović-Gjurić's own propensity for imaginative invention
is exemplified by her comments on the first Solvay Conference,
held in Brussels from October 29 to November 3, 1911, and
attended by the foremost European physical scientists, including Einstein. According to Trbuhović-Gjurić (1983, 97; 1988,
115), "during discussions [at the conference] Mileva attracted
great attention by her profound knowledge of scientific achievements and her clear understanding of their importance" In fact,

Marić did not accompany Einstein to the five-day conference, as two brief letters sent by Einstein to Marić during the outward journey testify (*CPAE* 5, docs. 300, 301).

MORE HEARSAY

Another example of an unlikely hearsay story concerns a visit by Marić's parents to the Einsteins' apartment in Zurich sometime in the early months of 1910. Trbuhović-Gjurić wrote that soon after the Einsteins had arrived back in Zurich from Bern in October 1909, "Mileva's home had become a meeting place for intelligent minds and talented musicians." She then reports that when Marić's parents returned to Novi Sad after their stay in Zurich in 1910, her mother proudly recounted:

> I didn't know that my Mitza [Mileva] was so highly esteemed in the world. When we were there, the most important and intelligent people came to their home and did not want to start talking until Mitza was present. Generally, she sat to the side and was content to listen, but as soon as she began to speak everyone turned to her and noted with great attention everything she said. (T-G 1983, 86; 1988, 104)

Trbuhović-Gjurić states that "Mrs Sofija Galić-Golubović of Belgrade, a cousin of Mileva's who grew up in the Marić family home, remembered well this report by Marić's mother." Krstić (2004, 141) also cites Golubović as the source (in July 1961) for his similar, though briefer, account of the parents' visit to Zurich in the spring of 1910. But in contrast to Trbuhović-Gjurić, Krstić suggests that the Einsteins' guests were most likely Paul and Conrad Habicht. At the beginning of March 1910, Einstein invited the brothers to stay in the Zurich apartment to finalize work on the small electrical device on which they had been collaborating (*CPAE* 5, doc. 198).

THE REJECTED DOWRY

Trbuhović-Gjurić and Krstić both recount yet another hearsay story, one that Marić's father Miloš supposedly told his son (also Miloš) and some of the latter's friends in Novi Sad. (In Krstić's version, the story was related to Miloš senior's "family and friends.") Miloš visited his daughter and son-in-law in Bern, and, according to Krstić, this occurred following the birth of their son Hans Albert in May 1904 (T-G 1983, 76, 1988, 94; Krstić 1991, 92). During this visit he had taken with him a bankbook for an account worth 100,000 Austro-Hungarian crowns and offered it to Einstein as a belated dowry. It was an offer, however, that Einstein refused in the following terms, as reported by Trbuhović-Gjurić:

> I didn't marry your daughter because of the money, but because I love her, because I need her, because we are both one. Everything I have done and accomplished I owe to Mileva. She is my genial source of inspiration, my protective angel against sins in life and even more so in science. Without her I would not have started my work, let alone finished it.

Trbuhović-Gjurić added that Marić completely agreed with the rejection of the dowry, and that Miloš wept when he told the young people in Novi Sad about it. However, the above account is suspect from the start. Einstein would not have said anything about having "finished" his work in 1904, when he had published only a few unimportant papers. Nor would he have said that without Marić he would not have started his work. There is also the improbability that the source of the story (see below) could remember in such detail the little speech purportedly uttered by Einstein after one week, let alone after several decades.

According to Trbuhović-Gjurić, the information came from a report by the journalist Miša Sretenović. Likewise, in his 1991

"Appendix," Krstić (1991, 92–93) states: "In a 1929 interview by Misha [Miša] Sretenović, Mileva stated that Albert had called her his inspiration, his guardian angel; someone who protected him from life's mistakes, but also from mistakes in his scientific work. This extraordinary gesture [Einstein's refusing to accept the money] was well remembered, and ever after, Albert was called 'our son-in-law' and was dear and popular in Novi Sad." Note that Krstić asserts that it was *Marić* who told the story to Sretenović, which we know from Marić's letter to Kaufler Savić in June 1929 was not the case (M-KS, 158). However, in his 2004 book he cites Marić's cousin Sofija Golubović for the story, stating in a footnote that she had told it to him in 1961 (Krstić 2004, 104, 104, n. 231). But if that was the case, why did he attribute the story to the journalist Sretenović in 1991 when he had a more direct source that he could name? Again, there is no evidence that Marić's father visited the Einsteins in Bern in 1904. All this raises grave doubts about whether the episode actually occurred.

In her 1990 article, Troemel-Ploetz prefaces the above quotation about the financial offer with the words: "He [Einstein] told Mileva Einstein-Marić's father" (T-P 1990, 418), as if it were a simple statement of fact rather than a story passed down through the generations. By the time it is reported by Peter Frize it has become something that Einstein "*wrote* to Mileva's father" (Frize 2009, 280, emphasis added). The historian Alberto Martínez notes that the same erroneous assertion is made in a replication of the quotation displayed inside the Einsteins' apartment at Kramgasse 49 in Bern, where the couple lived from 1903 to 1905.[9] Martínez observes: "There is no such letter to Einstein's father-in-law. The words are cited from a biography of Marić (from 1969) by Desanka Trbuhović-Gjurić, who claimed that Marić's relatives claimed to have heard Marić's father claim to have been told that by Einstein" (Martínez 2007).

A key to the mystery of the dowry may well lie in a discovery made by Michele Zackheim during her visit to Serbia in 1998. In the last home of Marić's mother, in a copy of *David Copperfield*, Zackheim found "a postal coupon that had probably been used as a bookmark." She continues (2000, 174): "It was a receipt for money, although the amount was not specified—and also for a five-hundred-gram package. It had been sent to Mileva, in Bern, from her father in April 1905" One might speculate that if it was the dowry, the postal coupon was sent by Miloš Marić to his daughter rather than directly to Einstein because, as far as is known, the only contact Miloš had had with his son-in-law before August 1905 was a letter he sent at the end of January 1902 to let Einstein know of the birth of baby Lieserl (Zackheim 2000, 43; *CPAE* 1, doc.134). He certainly would not have been well disposed toward a man who had brought shame on the Marić family name by making his unwed daughter pregnant, not to mention that Einstein did not make the effort to go to Novi Sad to see his fiancée despite her being so unwell after the birth that she couldn't write to him herself and had to leave the task to her father (*CPAE* 1, doc. 134). Nor, of course, did he and his wife receive an invitation to his daughter's wedding (Fölsing 1997, 106).

But, aside from illustrating the inconsistencies among versions of the supposed episode as reported by different authors, and the way the story mutated when other authors recycled it, all this turns out to be beside the point. Radmila Milentijević (2015, 114, 442, n. 252) has documented that the offer of a dowry was in fact *accepted* by Einstein. With this information added to the strong doubts about the supposed episode expressed above, and the discovery of a receipt for money *mailed* to Marić by her father in April 1905, there seems to be no good reason to believe that the episode actually occurred.

The publication of Desanka Trbuhović-Gjurić's biography of Mileva Einstein-Marić in Serbian (1969), and in German (1982ff) and French (1991) translations, followed by Senta Tro-emel-Ploetz's lengthy article in English in 1990 and (online) Evan Harris Walker's talk at the 1990 AAAS conference (Walker 1990), brought the fully developed "Mileva Story" to the public's attention. Despite the questionable nature of many of these authors' sources and claims, a number of widely read articles, books, and book chapters appeared over the next twenty-five years that uncritically repeated and even embellished elements of the story. The public was immediately enthralled, as it would be when such a story is suddenly brought to their attention, but its enthusiastic and unquestioning reception continued long af-ter the earliest appearance of these publications. While informed scholars judged the story almost entirely unfounded, its promo-tion continued unabated in print, plays, and television dramas, often with extravagant embellishments and misrepresentations of historical facts (see, e.g., Borchardt 2008, 2-3, and below).

A CONVERGENCE OF CULTURAL CHANGES

The reception of the Mileva Story raises the question why it was so readily accepted then and why it remains so, in whole or in part, among general audiences today. One answer is that the

story appeared during the convergence of several cultural
changes that are, in many ways, still with us today. That convergence lent instant plausibility to the story of a scientific wife in
the shadow of a highly celebrated scientist, her contributions
suppressed, her story nearly lost to history.

Growing out of the women's movement of the 1960s and
subsequent years, a burst of fresh scholarship on all aspects of
women in science began to emerge over the following decades,
including but not limited to biographies of women scientists,
studies of women's education and women's status in the scientific professions, critical assessments of gender roles in families
and dual-career marriages, and historiographic and philosophical critiques.[1] From many of these works a picture began to
emerge of women's struggles and strategies past and present for
equal education, for equal acceptance as scientists, for equal recognition for their work, and for the right to be remembered by
historians (Rossiter 1982, 1995). In this context, the story of the
hitherto practically unknown first wife of Einstein seemed to
coincide all too well with the stories of the lives of many other
women scientists and of their battles with male scientist colleagues or spouses, past and present. Among them was, for
example, the nuclear physicist Lise Meitner, whose colleague,
Otto Hahn, won a Nobel Prize for the discovery of fission that
should have been shared with her because of her important contributions to the discovery.[2] Rather than immediately acknowl-
edging her, writes her biographer Ruth Lewin Sime, Hahn
"quickly began to suppress and deny his collaboration with
Meitner" (Sime 1997; see elaboration in Sime 1996, chap. 14).
There were several others like Meitner, including some who *were*
recognized in their time, but who were since lost to history.
"What these women shared," writes Sime (1997), "was an

assiduous historical neglect, so that even the most famous were almost invisible to later generations."

The period circa 1990 also witnessed growing public skepticism regarding the authority and status of science and of prominent scientists—even a tendency toward debunking these and other elite figures, a tendency that continued and has reached new heights today in some quarters. In academe the "science wars" of the period, along with postmodern social-constructivist and related movements, provided an intellectual challenge to the traditional bases of scientific enquiry and its results that extended to scientists themselves. In 1987 Bruno Latour, one of the foremost members of this movement, declared that one of their aims was "to abolish the distinction between science and fiction" (Latour 1987, 166). More recently he stated, "I certainly was not anti-science, although I must admit it felt good to put scientists down a little. There was some juvenile enthusiasm in my style" (quoted in de Vrieze 2017).

Simultaneously with these developments came the publication of the first volume of the *Collected Papers of Albert Einstein*, perceived by some as a renewed celebration of Einstein, which in turn reinvigorated the small number of those who continued the long-standing anti-Einstein tradition that derived from the early 1920s in Germany and spread across national, cultural, and temporal boundaries. Over the past century their numbers have included anti-modernists, anti-Semites, anti-relativity theory proponents, as well as demagogues representing themselves as defenders of culture and religion against a supposed "relativism" conflated with Einstein's relativity (Hentschel 1990; Cassidy 2004; Wazeck 2014). An instance illustrating an anti-Einstein sentiment came in a statement quoted by the *New York Times* in which Evan Harris Walker revealed that his purpose in supporting the Mileva Story was not so much to bring long overdue recognition to Marić, but rather to undercut Einstein's reputation: "My point is to say that the king has no clothes. I'm not saying that Albert didn't do anything, but because she was older and was initially the leader, she was probably the source of some of the ideas" (*New York Times* 1990).

Within this fervent mix circa 1990, the sudden appearance on the public stage of Einstein's little-known and reportedly unjustly treated first wife, as portrayed by Trbuhović-Gjurić and publicized by Troemel-Ploetz and Walker, made Marić an immediate exemplar of much that was and is wrong with male-dominated science and scientific culture, with historical research that had neglected her and many others all those years (and still does), and with Einstein himself. Troemel-Ploetz made these features explicit for anyone who missed them: "Trbuhović-Gjurić's motivation was to focus on the unknown, unacknowledged, and on what was 'unjustly put aside into oblivion.'" A little later she reiterated: "[Trbuhović-Gjurić] wanted to rescue Mileva Einstein-Marić from oblivion and write her into Serbian and scientific history. She knew that no man would do that job for Mileva Einstein-Marić, whose own husband failed to give her the public recognition she deserved." (T-P 1990, 417)

The match between the Mileva Story as portrayed by the authors cited above on the one hand, and the scholarly feminist critiques, skepticism about science, and some suggestive sentences in the newly published early Einstein-Marić letters, on the other, appeared so complete that many writers and journalists accepted it seemingly without question and promoted it without careful investigation of its sources. It was a reaction closely related to confirmation bias, the ready acceptance and tendentious interpretation of evidence as confirmation of a pre-existing belief or opinion (Plous 1993).

In the following sections I examine (in chronological order) how some representative authors in print and other media have handled the Mileva Story to the present day. My treatment is not exhaustive (however exhausting it may be); most authors have tended to repeat uncritically the same few arguments and claims presented over the years by their predecessors. Some repetition is inevitable, however, especially when familiar arguments are supplemented and enhanced.

ANDREA GABOR, *EINSTEIN'S WIFE* (BOOK, 1995)

One of the most extensive early secondary accounts of the Mileva Story for a broad audience is found in "Mileva Marić," the first chapter of Andrea Gabor's 1995 book *Einstein's Wife: Work and Marriage in the Lives of Five Great Twentieth-Century Woman* (Gabor 1995, xi–xiii, 1–32). The book was widely reviewed.[3] Gabor's account in her chapter on Marić repeats many of the assertions made by Trbuhović-Gjurić without questioning their validity, or even their plausibility. The errors and misconceptions are so numerous that they justify the description of many passages in the chapter as "flights of journalistic fantasy" (Schulmann and Holton 1995).

Many of the misleading contentions that Gabor recycles from Trbuhović-Gjurić and Krstić, such as the Joffe Story and those relating to the Milana Stefanović interview of 1929 (Gabor 1995, 20–21), have been discussed at length in previous chapters. Rather than working through them again, I focus below mainly on aspects of the story surrounding the Polytechnic years that have not been previously discussed or that bear further clarification.

ENTRANCE EXAMS

Although central European students who had passed the Matura exams at the end of their gymnasium (high school) studies were

generally admitted to higher education without further examination, both Marić and Einstein took admission exams to the
Swiss Polytechnic Institute (Zurich Polytechnic). Gabor (1995,
15) contrasts their performances by asserting that Marić "passed
the difficult ETH entry examination (which Einstein had failed
the first time he took it)." This statement is grossly misleading.
With the exception of one subject, Marić did not even take the
Polytechnic entrance exams, and in any case their situations were
entirely different. Since Marić had passed the Matura exams at a
medical school after leaving the Zurich Higher Girl's School, she
should have been admitted automatically to the Polytechnic
without further exams. But for some unknown reason the Polytechnic required her to take the mathematics components of the
entrance exams. The first page of her Polytechnic student report
(Matrikel), published by Trbuhović-Gjurić (T-G 1983, 56;
1988, 60) and now available online (Marić, Student Record),
listed her grades on these exams. Marić obtained a very moderate
grade average of 4.25 on a scale 1 to 6 (see appendix A).

Einstein was required to take the full Polytechnic entrance
exams because he had left his Munich high school two and a half
years early, well before he could take the Matura exams (called
the Abitur in Germany). Gabor fails to mention that Einstein
was still only sixteen at the time he took the entrance exams
(Marić was twenty), and that he had been out of the school system for some nine months while living with his parents in Italy.[4]
His grades on the exams in physics and mathematics were excellent, but he failed the exams overall because of his low grades in
the other subjects (Fölsing 1997, 37; *CPAE* 1, p. 11 and n. 7).
He subsequently attended the nearby Aargau Cantonal School
in Aarau to complete his secondary education. There his grades
in physics and mathematics were again excellent, with maximum grade 6 in arithmetic/algebra and geometry (appendix B).

When he sat for the Matura exams at the end of that year he passed, achieving, according to documents cited by the Einstein editors, the highest average grade (5⅓) of the nine students in his class who sat for the Matura exams although he was the youngest of the candidates (*CPAE* 1, ed. note, p. 25). Now seventeen years old, Einstein entered the Polytechnic in fall 1896 without being required to take the entrance exams again (as Gabor mistakenly implies). Being one year below the official minimum age for admission, he was the youngest of six students enrolling in 1896 for a diploma for teaching mathematics and physics in secondary schools (Seelig 1956a, 24).

INTERMEDIATE AND FINAL DIPLOMA EXAMS
Marić took her intermediate diploma examinations in 1899. This was one year later than Einstein and the other students in her class cohort because of her semester away at Heidelberg University. Gabor writes:

> Despite missing one semester, she passed her first year's examinations, her highest grade 5.5 out of a possible 6. By contrast, Einstein's final grade in physics was 5.25, and he received his only 6 in electrical engineering. (Gabor 1995, 10)

The words and figures in this passage are both confusing and misleading, not least because Gabor does not compare like with like. First of all, Marić did not miss one semester in her first year (it was the first semester of her *second* year). There were no first year examinations and Marić did not achieve more than 4.5 in any of the topics (all mathematics plus mechanics) for the course grades for the two semesters of her first year (appendix C). (Einstein's first year semester grades were not much better, though he achieved 5 in two of the topics.) The only 5.5 that Marić achieved while at the Polytechnic was for physics in the intermediate diploma exams in 1899 (appendix D), so it is this grade to which Gabor must be alluding. On the corresponding examinations a year earlier, Einstein had also obtained 5.5 in physics, but whereas he came out the first of five candidates with an overall grade average on all of the exams of 5.7, Marić's overall grade average was 5.05 and she came fifth of six candidates (Zimmermann 1988a, 63).

The "final grade in physics" of 5.25 that Gabor cites for Einstein is his physics grade on his "Leaving Certificate" at the end of his Polytechnic studies (appendix E). By comparison, Marić's corresponding physics grade, not provided by Gabor, was 5, as shown on a facsimile of her leaving certificate in Trbuhović-Gjurić's book (T-G 1983, 57; 1988, 61; see appendix E). Presumably this was an unspecified average of semester physics grades. As Gabor writes, Einstein received the only 6 on his leaving certificate in Electrotechnical Laboratory, the same grade he received in that course during his third year. No grade was given for Marić for that subject on her certificate, presumably because she did not take it. In their third year, however, both students received a 5 for Scientific Projects in Physics Laboratories, and 6 and 5 for that subject for the two semesters in their fourth year (appendix C).

A comparison of their semester grades throughout the four-year course, as recorded on their respective final leaving certificates (appendix E), shows that out of the nine subjects they took in common, Einstein's grades were higher in five, and they were equal to hers in two. The two for which Marić's grades were higher than Einstein's were Determination of Geographical Location and Introduction to Physics Practical for Beginners. For the latter course Einstein was awarded the lowest possible grade of 1, owing to his poor attendance record, for which he received an official reprimand (*CPAE* 1, doc. 28, p. 47 n. 4;

Fölsing 1997, 57). (Einstein skipped classes for this course because of conflicts he had with the professor [Jean Pernet] taking the course [Seelig 1956a, 40–41].)

DISCRIMINATION

The overall grades for the final diploma examinations were composed of a weighted average of the subject grades and the (heavily weighted) grade on an experimental research thesis. Marić unfortunately failed twice to pass the final diploma exams, the second time, in 1901, under great personal stress, the most notable of which being that she was some three months pregnant. Gabor suggests her failure on the first (1900) diploma exams might have arisen from discrimination in the Polytechnic's treatment of her in the context of another misleading characterization of her grades:

> Although the final grades for both Marić and Einstein fell below the 5 point average that was necessary to pass, Einstein's 4.9 [actually 4.91] got rounded up to 5 so that he just barely squeaked by. Marić's 4, on the other hand, meant that she failed outright; once again, she had gotten high marks in physics, but it was a miserable 2.5 average in the Theory of Functions that dragged down her final grade. (Gabor 1995, 13–14)

Gabor's assertion that Marić obtained "high marks in physics" could apply perhaps to her exam grade of 5 for Experimental Physics, but in theoretical physics she achieved only 4.5 (see appendix D, some grades were doubled). Gabor provides no source for her contention that an average grade of 5 was the passing mark for the diploma grade average, and that Einstein's grade was "rounded up" to enable him to pass. No criteria for passing are cited in the official examiners' reports (*CPAE* 1, doc. 67; Minkowski 1901), and none could be found in the official examination regulations at the Polytechnic, now ETH (Stachel 2002, 32). It was apparently a committee decision made on an individual basis. However, at the 1990 AAAS conference Troemel-Ploetz made the assertion that Einstein's diploma grade was "rounded up to 5" to enable him to obtain his diploma (Stachel 2002, 32), a claim she repeated in the 2003 TV film *Einstein's Wife: The Life of Mileva Marić Einstein* (Hilton 2003; see below) in which Gabor appeared and where she no doubt obtained her (erroneous) information.

Gabor (1995, 15) writes, without a reference citation, that the Einstein historian Robert Schulmann "suggests that because the finals included an oral component, she might have been subject to the prejudice of her examiners." (Gabor 1995, 13–14) While the possibility of discrimination cannot be ruled out, it should be noted that Marić's grades in the five intermediate diploma exams in 1899, also oral, were all 5 or above, with the exception of Descriptive Geometry & Geometry of Position, for which she achieved 4.75. (Her difficulties with descriptive geometry, the professor for which subject was Wilhelm Fiedler, were acknowledged by Marić herself in a letter to Einstein in late summer 1899 [E-M, 12].) While no one doubts the obstacles faced by women students aiming for university-level credentials in science subjects even in relatively enlightened Switzerland, Stachel (2002, 29–30) has documented from the ETH archives that about one-sixth of the students in Section VI for intending teachers of science and mathematics in secondary schools at that time were women. While this is a low proportion, it compares very favorably with other countries in Europe at the end of the nineteenth century, many of which would not have allowed female students even to enter higher level educational institutions. This certainly does not rule out the possibility of prejudice against Marić by one or more of the final diploma examiners, but nor is there hard evidence that there was such prejudice.

On this same issue, Troemel-Ploetz (1990, 422) makes a much more emphatic charge of discrimination against Marić on the part of "the elitist male ETH." After observing that "[w]e cannot imagine what the atmosphere must have been like for Mileva Einstein-Marić when she came to study there in 1896," she claims that "[t]he general attitude [at ETH] was, and is, that women do not belong there, so there are no positive expectations for them in the heads of their male professors ... I do not believe that even the physics professor, Weber, for whom she wrote her Diplomarbeit [diploma thesis], which she wanted to extend into a doctoral dissertation, thought about taking her on as her assistant." As noted above, Stachel has documented that Troemel-Ploetz's characterization of the Polytechnic (ETH) as uniformly antagonistic to women at the end of the nineteenth century is not an accurate description of that institution. More specifically, it is not the case that Professor Weber did not think of employing Marić as an assistant, as Troemel-Ploetz surmises (M-KS, 61; see below).

LOST ASSISTANTSHIPS

On another issue Gabor relates Einstein's supposedly mediocre academic performance to his difficulty in finding a job after graduation:

> Einstein was, in fact, a mediocre student and one of the few ETH graduates who did not receive a position as an assistant upon graduation, which made it difficult for him to obtain a job. (Gabor 1995, 12)

It is true that, although Einstein was top of his group in the 1898 intermediate diploma examinations, during the final two years he was inclined to neglect his coursework, concentrating instead on his own private research on advanced physics that was absent from Weber's course. In his later "Autobiographical Notes," Einstein recalled those days: "I followed several lecture courses with great interest, but otherwise I 'cut' classes often and studied at home the masters of theoretical physics with holy zeal" (Einstein 1955). Frequent references to new theoretical ideas, and to outside readings of advanced physics texts, alone and together with Marić, can be found in the Einstein-Marić correspondence (E-M).

Given such circumstances, Einstein's final average diploma grade of 4.91 in 1900 was quite creditable. In the four subject examinations he achieved at least 5—it was his heavily weighted grade for his dissertation that reduced his average grade to below 5. It was primarily Einstein's poor relationship with the two physics professors, Heinrich Weber and Jean Pernet (Fölsing 1997, 57, 72), that accounts for his failing to obtain an assistantship at the Polytechnic, though his mediocre 4.5 for his diploma dissertation may have also have been a factor. His failure to obtain an academic appointment elsewhere he blamed on Weber for not giving him a recommendation (E-M, 36). That he had not obtained a permanent post of any kind by December 1901 led Marić to suggest to Kaufler Savić that his "very wicked tongue" and the fact that "he is a Jew" was at the root of the problem (M-KS, 79).

In contrast to Einstein, in her Introduction Gabor (1995, xiii) writes that Marić enjoyed "an easier relationship with their professors," and, because of Einstein's difficulties with them, she "often interceded with them on Einstein's behalf." In her chapter on Marić, Gabor suggests that by doing so she "may have damaged her own academic relationships," and reiterates this later in the same paragraph:

What is clear is that Marić tried several times to intercede with Weber on Einstein's behalf—an effort in which she was unsuccessful and one that may have eroded her relationship with the professor. In the summer of 1901, Mileva wrote to her friend Helene Kaufler Savić: "I've already quarreled with Professor Weber two or three times, but now I am already used to such things. Because of him I have suffered a lot. … We still do not know what destiny has determined for us [Albert and Mileva]. (Gabor 1995, 13, her ellipses and insert)

Gabor evidently based this passage on Krstić's evidence-free contention that "Mileva tried to persuade Weber to take Einstein as an assistant, but the professor was adamant" (Krstić 1991, 90; Gabor 1995, 295), though by the time she recycled the story it had become that Marić interceded on Einstein's behalf "several times." Note that Gabor asserts that it is "clear" that what she wrote was the case, a terminology that serves to persuade the reader that there is little doubt that it is historical fact. Note also that Gabor's translations (above) of the three sentences from Marić to Kaufler Savić are identical to those in Krstić (1991, 90), which publication she cites in the relevant references at the back of her book, and that the way they are presented gives the appearance of their being from a single letter written "[i]n the summer of 1901." In reality, however, they are actually from three *different* letters. Moreover, both Gabor and Krstić omitted the ellipses between the first and second quoted sentences, making it appear even more definite that they came from a single letter. But only the first sentence was editorially attributed to summer 1901 ("[May-June, 1901]" in M-KS, 73, 76): the other sentences are in two different letters from Marić to Kaufler Savić sent respectively several months after and before summer 1901 (M-KS, 78, 72). If the three sentences quoted by Gabor (following Krstić) are read in their actual contexts it can be seen that they have no bearing on any purported interventions by Marić on Einstein's behalf (see below), and one of them shows explicitly that her arguments with Weber pertain to his criticisms of her doctoral dissertation (M-KS, 78; see also E-M, second half of a letter dated May? 1901, 51–52).

WEBER'S CRITICAL COMMENTS

As noted above, the main source of Marić's problems with Weber in 1901 appears to have been in relation to her doctoral dissertation on which he was her advisor, rather than with Einstein's strained relationship with him. Marić had completed her diploma thesis in July 1900 and planned to continue that research for a doctorate (E-M, 30, 32). Einstein wrote to her in May 1901 regarding her research: "So, how is your work going, sweetheart? Everything going well in your life? Is old Weber behaving decently, or does he again have 'critical comments'?" (E-M, 51–52). About that time, in May or June 1901, Marić reported to Kaufler Savić that she had had "several arguments with Weber, but I am used to it by now" (M-KS 76). Then in fall 1901, after failing the diploma exams for the second time, Marić reported to her friend: "I have finished my studies, although thanks to Weber's concerns [*Fürsorge*], I have not yet managed to obtain a doctorate. I have put up with a lot from him and will on no account go back to him again" (M-KS, 78; Popović 1998, 187). Note that Weber's "concerns" are the reason given for her discontinuing with her doctoral dissertation, which evidently relate to what Weber saw as shortcomings in her research project, rather than to any conflict regarding Einstein.

Aside from the numerous factual errors highlighted above, Gabor's chapter takes as given Trbuhović-Gjurić's claim that Marić played a major role in Einstein's early researches, most notably in relation to the 1905 papers. To take just one revealing example, she asserts that "[d]uring the early years of their marriage, Marić also spoke freely to her family and friends about collaborating with her husband" (Gabor 1995, 20–21). In support of this broad generalization Gabor cites two items of evidence, both of which are suspect. She writes, referencing Trbuhović-Gjurić (Gabor 1995, 295), that "Marić told Milana Bota [Stefanović], for example, about the work she was doing with Einstein." For her second example she writes, citing Krstić's 1991 "Appendix A": "Marić boasted to her father and Desana Tapavica Bala … '[j]ust before we left for Novi Sad, we finished some important work that will make my husband world famous.'" These items have been closely examined in chapters 8 and 5 respectively, and found to be completely unreliable as historical evidence. Gabor, however, accepts Trbuhović-Gjurić's and Krstić's reports without caveat.

GERALDINE HILTON, *EINSTEIN'S WIFE* (TELEVISED FILM, 2003)

The most widely disseminated account of the Mileva Story came through the televised film *Einstein's Wife: The Life of Mileva Marić Einstein* (Hilton 2003), written by Geraldine Hilton and directed by Nicola Woolmington, in association with Oregon Public Broadcasting in the United States and the Australian Broadcasting Corporation (ABC). Both ABC and the Public Broadcasting Service (PBS) organizations in different states in America aired the film, which was a runner-up for the 2004 Japan Prize for Educational Media in the Adult Education division. An educational website was set up to accompany the program; the extensive, downloadable, lesson plans for secondary school teachers included tendentiously worded questions for their students that implicitly accepted everything in the film as factual history.

Trbuhović-Gjurić's version of the Joffe Story took a prominent place in the film as alleged conclusive evidence that the names of both Einstein and Marić were on the original manuscripts of the 1905 papers. The accompanying PBS website even described Joffe as "among her [Marić's] supporters," contending that "there is at least one printed report in which Joffe declared that he personally saw the names of two authors on the 1905 papers" (the source is no longer online). However, as the historian Alberto Martínez writes:

> In a particularly careless confusion, the producers of *Einstein's Wife* and the companion website pictured a fragment of a page that reads "signed Einstein-Marity," purportedly written by Joffe. But the page pictured is instead from a popular science book from 1962, by a Russian writer Daniil Semenovich Danin, who again did not even claim to have ever seen the original *manuscripts* or to have known anyone who had [ref.: Danin 1962, 57]. (Martínez 2005, 52; emphasis in original)

A synopsis of the film published on the Japan Prize website illustrates the intentions of the film's writer and director and reveals the sensationalism with which it was promoted:

> In a shocking discovery that would rock the scientific community, a secret cache of love letters written by Albert Einstein to a first wife were uncovered in the Einstein archives in 1986, over 30 years after his death. Presented to the public by physicist Dr. Evan Harris Walker at a North American science convention in 1992, these letters would reveal a passionate and highly collaborative relationship between Albert Einstein and Mileva Marić and raise questions regarding Einstein's sole credit for his Theory of Special Relativity. It was this event that would lead producer Geraldine Hilton on an eight-year odyssey, researching the life of this unknown scientist. *Einstein's Wife* reveals the truth behind one of the greatest scientific collaborations of the 20th century. Rescued

from historical obscurity, the story of Mileva Marić is finally being told. *Einstein's Wife* educates its viewers not only on the life and scientific accomplishments of Mileva Marić, but it also challenges its viewers to examine the role of women in the late nineteenth and early twentieth centuries.[5]

Regardless of who wrote the above summary, eight years of research should have made clear that the letters were not uncovered in the Einstein archives but, more precisely, among Hans Albert's family papers in California in 1986 (E-M, xiii–xiv; Highfield and Carter 1993, 6, 278–281). The letters were first presented to the public, not by Evan Harris Walker in 1992, but in the first volume of *The Collected Papers of Albert Einstein* in 1987, when they were published both in the original German and in English translation (*CPAE* 1 and *CPAE* 1, Engl.). Shawn Smith's more polished English translation appeared in 1992 (E-M). Although these early letters reveal Einstein's extraordinarily precocious ideas on the frontiers of contemporary physics, there is no substantive evidence in them that there was a "highly collaborative [scientific] relationship" between Einstein and Marić. What they actually show is that Einstein had (unrealized) hopes of such a relationship when they would eventually marry, and that he needed someone to whom he could pour out his latest ideas, despite the fact that the surviving letters from Marić show that even when he enthusiastically reported to her his latest ideas in some detail, they evoked no response from her in the direct replies she sent to Einstein (E-M, 10–11 and 12–13; 47–48 and 49). (See chapter 7.) More specifically, they do not "raise questions regarding Einstein's sole credit for his Special Theory of Relativity" as claimed in the above synopsis of the film. But that was the story broadcast across the United States by PBS, and across Australia by ABC, in 2003.

Following my detailed complaint to Oregon Public Broadcasting regarding the film and the accompanying website (Esterson 2006a and 2006b), along with supporting statements from
Gerald Holton, Robert Schulmann, and John Stachel, on
December 15, 2006, the PBS Ombudsman Michael Getler
posted what he described as an "exceedingly long column"
upholding the complaint, in which he also included Hilton's
responses to it. Getler concluded that the PBS production was a
"factually flawed and ultimately misleading combination of film
and web presentations" (Getler 2006). Although Getler advised
PBS to cease selling the DVD of the film pending a review by a
small group of scholars and historians of science, they commissioned instead Andrea Gabor, despite her strong partiality, to
revise the website material.[6]

Reconsidering her position on some of the issues, Gabor
made considerable corrections to the website, but, as a testa-
ment to the persistence of repeated publications of error, the
revised version of the website, posted in 2007 but since "retired"
by PBS, still contained several errors and misconceptions. These
included a partially revised statement on a central issue, the Joffe
Story, that retained major errors as indicated here by the high-
lighting of erroneous phrases through the use of italics:

> [Joffe] *saw the original version* of Einstein's three most famous pa-
> pers (on Brownian Motion, the photoelectric effect, and the theory
> of relativity) and *said they were signed Einstein-Marity*, Marity being
> the Hungarianized version of Marić. *Whether that reference was to
> one author or two is the nub of the debate.* Most researchers now
> agree, based on a memorial to Einstein, written by Joffe, that he was
> referring to a single author rather than a husband-wife team.

Gabor's idea as to what is "the nub of the matter" is mistaken.
That actually centers on Trbuhović-Gjurić's whole passage,

which is completely misleading but taken as historical fact by purveyors of the Mileva Story (T-G 1983, 79; 1988, 97; see chapter 8). However, despite the substantial revision of the website material, PBS eventually decided to cease broadcasting the film, and to take down the website.

In contrast to PBS's equivocations, in response to my complaint the Australian Broadcasting Corporation acknowledged breaches of the ABC's Code of Practice and made the decision not to broadcast "Einstein's Wife" again, and to remove its school study guide from the ABC website.

DORD KRSTIĆ, *MILEVA & ALBERT EINSTEIN* (BOOK, 2004)

In his book *Mileva & Albert Einstein: Their Love and Scientific Collaboration*, Dord Krstić fully supports the Mileva Story as Trbuhović-Gjurić and Troemel-Ploetz present it, and he even goes beyond them by claiming that the couple "regularly collaborated together on joint scientific work, for approximately fifteen years (1898 to 1913)." In the same section he alludes to "eyewitnesses" who he claims to have "spoken and written independently about this teamwork," and to what he calls "scientific manuscripts" in Marić's handwriting from around 1909/1910 that he has analyzed in his book. After adding that "[i]n the future some additional proof may also be found," he writes: "Therefore, it can be concluded that a majority of articles from that period (for which officially only Albert was stated as the author), were the result of the couple's collaborative scientific work. Keeping in mind today's criteria for authorship credit given on scientific papers, Mileva, one of the first female theoretical physicists in the world, was the co-author of those important articles." (Krstić 2004, 204)

MANUSCRIPTS IN MARIĆ'S HANDWRITING
I examine below the evidence that, according to Krstić, confirms
the couple's scientific collaboration over some fifteen years,
starting with the material in Marić's handwriting.

THE PLANCK LETTER   Volume 3 of Einstein's *Collected Papers*, published in 1993, contains an undated document, described by the editors as being "in the hand of Mileva Einstein-Marić," with the heading "Antwort auf Planks [*sic*] Manuskript" (Response to Plank's Manuscript) (*CPAE* 3, doc. 3). It is the unsigned draft of a letter addressed to the eminent physicist Max Planck containing remarks on a pre-publication manuscript that Planck evidently sent to Einstein for comment. The final version of Planck's manuscript was received by the *Annalen der Physik* on January 18, 1910, and published the same year (Planck 1910), so the document must have been written some time prior to that date.

Krstić (2004, 149) makes much of this document: "It is very significant that the Einstein answer to Planck's article is in Mileva's handwriting. It is more than symbolic; it proves Mileva's active and important role in Albert's scientific work." In order to judge whether Krstić is justified in drawing this conclusion it is essential to look closely at both the content of the letter and the background context that can be ascertained from some knowledge of the theoretical ideas it contains. The background may be determined from Einstein's correspondence during the previous year. In a letter to the physicist Jakob Laub in May 1909, Einstein reported his intense involvement with the theoretical notions that appeared in the draft letter, and his differences with Planck:

> I busy myself incessantly with the question of the constitution of
> radiation, and am conducting a wide-ranging correspondence
> about this question with H. A. Lorentz and Planck. The former is

an amazingly profound and at the same time lovable man. Planck is also very pleasant in his letters. His only failing is that it is hard for him to follow other people's trains of thought. This might explain how he could have made totally wrong-headed objections against my last radiation paper. But he did not adduce anything against my criticism. So I hope that he has read it and accepted it. This quantum question is so extraordinarily important and difficult that everybody should take the trouble to work on it. I already succeeded in thinking up something that is formally more or less adequate, but I have good reasons for regarding it as "garbage" nonetheless. (*CPAE* 5, Engl., doc. 160)

In a lengthy letter to Lorentz in the same month, Einstein discussed the details of his researches following the publication of his paper on fluctuations in March 1909 (*CPAE* 2, doc. 56; *CPAE* 5, doc. 163). In it he presented his own new theoretical ideas as well as his differences with Lorentz; and in a letter to his friend Michele Angelo Besso at the end of 1909 he wrote: "I haven't discovered much. The most interesting discovery is that one can specify an infinite manifold of energy distributions that are compatible with Maxwell's equations. Perhaps herein lies the solution of the quantum question." He then went on to give the details (*CPAE* 5, doc. 195). This "discovery" is one of the items presented in the draft letter to Planck.

At the end of Krstić's book the historian of physics Stanislav Južnić provides a commentary (Krstić 2004, appendix C, 236–244) on the draft letter in which he notes, as indicated above, that Einstein had been working during 1909 on ideas related to the subject matter of Planck's draft paper (see also *CPAE* 3, ed. preface, xviii–xxi). He then goes on to provide the following details about the draft letter:

Einstein's notebook with a copy of the letter written in Mileva's handwriting, survived on two lined pages, along with a simple geometric picture. Mileva crossed out several mistakes, but left some unnoticed errors. The manuscript was probably the dictated first draft, which was later corrected and mailed to Planck. Among other errors, the addressee was written as "Plank." It seems almost impossible to imagine Einstein sending to the imminent [*sic*] Planck a letter with such a huge error, almost an insult. Mileva's spelling of the last name was not unusual, because Josef Plank was a former assistant of Josef Stefan [both physicists] in Vienna ... Einstein criticized the 6th page of Planck's manuscript. Mileva crossed out the letter "*d*," probably as the beginning of the wrongly used indefinite article between the words "*Zahl*" and "*quantenhaftger*" [*sic*]. She crossed out the word, "*kann*" (can) and replaced it with "*müsse*" (must) at the very end of the letter, with no period to finish the sentence with. ... In the middle of the letter, Mileva crossed out the word "*besteht* (valid)" and replaced it with the more defined "*lässt*" for Einstein's and Ludwig Hopf's (1884–1939) new theory of radiation. (Južnić 2004, 240–241)

As the listed errors suggest, the document in Marić's handwriting bears the hallmarks of a "dictated first draft" of Einstein's response to Planck's draft paper. This is also implicit in Južnić's remark that "Einstein criticized the 6th page of Planck's manuscript." Only someone totally immersed in the subject matter, as was Einstein, would be in a position to comment on Planck's paper. The complexity of the subject matter is demonstrated by his correspondence with other eminent physicists, for example his lengthy exchanges with Hendrik A. Lorentz (*CPAE* 5, docs. 153, 163; see also Irons [2004]). There is no evidence that Marić had been so immersed, and in fact Einstein's total dedication to his scientific work inevitably led to her being neglected, as the following passage from a letter Marić wrote to Kaufler Savić in winter 1909–10 indicates:

> [My husband] is working very much, and he has published a great deal, and is now trying to learn about the practical side of physics,

for which he has not had much opportunity. ... You see, with that
kind of fame he does not have much time for his wife. I read between the lines [of Kaufler Savić's letter] a certain impish tone when
you wrote that I must be jealous of science. But what can you do?
One gets the pearl, another the box. ... You see, I long for love, and
I would so rejoice if I could hear an affirmative reply that I almost
believe it is the fault of the damned science, so I gladly accept your
smile on that account. (M-KS, 101–102)

It is inconceivable that Marić would have written in such
terms if she had been playing an "active and important role in
Albert's scientific work" at this time, which is what Krstić (2004,
148–149) claims the draft letter to Planck proves.

Milentijević (2015, 144; 445, n. 335), citing Krstić's 2004
book, takes a similar view of the draft letter, writing that it "confirms how closely Mileva worked with Albert." However, she
fails to mention the several mistakes crossed out in the draft, and
other unnoticed errors, though the fact that she used material in
Krstić's 2004 book on several occasions makes it scarcely credible that she failed to read Appendix C, listed on the Contents
page as "Dr. Stanislav Južnič: 'Answer to Planck written in Mileva Einstein's Handwriting'" (Krstić 2004, 236–244).

SEVEN PAGES OF NOTES AND ONE DIAGRAM Document 3 in
the third volume of the *Collected Papers of Albert Einstein* comprises Einstein's lecture notes, contained in two notebooks, for
the course in introductory mechanics that he taught at the University of Zurich in 1909–10. An editorial note (*CPAE* 3, p.
125) records that at the back of the second notebook there were
"seven pages of notes in Mileva Einstein-Marić's handwriting,
containing material very closely corresponding to the introductory section of the first notebook, followed by an eighth page
with a drawing of three intersecting circles, also in Mileva

Einstein-Marić's hand." Immediately prior to discussing these seven pages, Krstić (2004, 142) writes, following Trbuhović-Gjurić, that Svetozar Varičak's daughter had recalled that her father, who had lodged with the Einstein's in Zurich when he was a student, recounted to her that "Mileva, after working hard all day, would at times stay up past midnight to solve mathematical problems in Albert's notes." (See chapter 8.) Krstić continues:

> The proof of Svetozar's observations can be found in Albert's lecture notes for his lectures on mechanics for the academic year 1909/10 (the second notebook), which were saved in the original handwritings, and are kept in "The Albert Einstein Archives, The Hebrew University of Jerusalem, Israel." Namely, *eight (7 +1) pages of those notes from analytical mechanics are in Mileva Einstein's handwriting.* These notes from analytical mechanics are historic, and they prove that at that time Mileva was active in the study of physics and that she collaborated with Einstein. (Krstić 2004, 142, emphasis in original, refs. omitted)

Krstić adds:

> I personally checked the authenticity of Mileva's handwriting; and the fact that these eight pages in Albert's notebook are in Mileva's handwriting is also documented in the footnote of [ref in original: *CPAE* 3, doc. 1, p. 125]. ... The first page of Mileva's manuscript about analytical mechanics in Albert's notebook is published for the first time in this very book [photograph, p. 143]. ... They are evidence of her creative process and knowledge, for the corrections on all seven pages are in her handwriting. (Krstić 2004, 144)

There is a confusing element relating to the seven pages of notes on mechanics in Marić's handwriting. Their title in Krstić's photograph reads "Analytische Mechanik," a course that Einstein taught at the ETH during the winter semester of 1912–13, when, as discussed in chapter 8, earlier, Svetozar Varičak was staying at the Einstein home (*CPAE* 3, appendix B). But the notes themselves correspond to a course titled "Introduction to Mechanics," which Einstein taught during the winter semester 1909–1910 at the University of Zurich, his first semester teaching at that university (*CPAE* 3, 125, ed. note). This is supported by a comparison of the photograph of the first page provided by Krstić with the early pages of Einstein's first lecture notebook for that course, published as *CPAE* 3, doc. 1. Of course, it may be that she simply wrote the wrong title on her notes.

That Einstein took care in his preparation of lecture notes for the winter semester 1909–10 is evident from his writing to Lucien Chavan in December 1909: "I get much pleasure from teaching, even if it requires much work the first time around." Later that month he told Jakob Laub: "I take my lectures very seriously, so that I must spend much time on their preparation" (*CPAE* 5, docs. 193, 196). Einstein makes no mention in these letters of receiving any help on this task from his wife, not even to Laub, who knew Marić quite well because he had stayed with the Einsteins the previous year for at least five weeks, collaborating with Einstein on two papers (*CPAE* 5, docs. 91, 96; Fölsing 1997, 240, 240, n. 25; *CPAE* 2, docs. 51, 52). Nor, when Marić tells Kaufler Savić in June 1909 that "[m]y husband is now very busy preparing his lectures," does she report that she has been assisting him in this task (M-KS, 97). Nevertheless, after briefly repeating the account from his cited source (Trbuhović-Gjurić) of what Svetozar Varičak's daughter was reported to have said, Krstić (2004, 142) asserts that Marić "spent many hours working on these [1909/1910] lectures," evidence for which he doesn't provide.

It is evident that we are not in a position to account for the seven pages of lecture notes in Marić's handwriting, closely corresponding to the early pages of Einstein's own notes found in his second notebook. Krstić (2004, 144) contends that "[t]hey are evidence of her creative process and knowledge, for the corrections on all seven pages are in her handwriting." Moreover, he maintains earlier in his book that these pages, together with the unsigned draft letter to Planck, "are important complementary evidence about the scientific collaboration between Mileva and Albert Einstein" (Krstić 2004, 15). However, as we have seen, these assertions do not withstand even a cursory examination. In regard to Krstić first claim, the seven pages of lecture notes on introductory mechanics obviously do not reveal anything about Marić's knowledge of physics beyond university level. In regard to the draft letter to Planck, it is inconceivable that at that time she was immersed in the level of theoretical physics necessary for taking part in the exchanges between Einstein and Planck. Furthermore, in 1912, around the time that we know with some certainty that Svetozar Varičak was lodging with the Einsteins, Marić reported to Kaufler Savić (M-KS, 107–108): "My big Albert has become a famous physicist who is highly esteemed by the professionals enthusing about him. He is tirelessly working on his problems: one can say he lives only for them. I must confess with a bit of shame that we are unimportant to him and take second place." These words of Marić's are not compatible with Krstić's presumption, expressed elsewhere in his book, that even up to 1912 she was collaborating with Einstein on his researches in theoretical physics (Krstić 2004, 157).

ALLEGED OBSERVATIONS OF COLLABORATION
Krstić (2004, 204) makes his most extravagant claim when he writes: "There is much evidence, some of it published here for the first time, which confirms that Albert and Mileva … regularly collaborated together on joint scientific work, for approximately fifteen years (1898 to 1913)." From this he contends that

Marić was co-author of twenty-two of Einstein's published papers, with a further seven for which her collaboration is "uncertain" (Krstić 2004, 221, n. 472, 221–225). As we shall see, these claims are not based on any solid evidence but instead depend on unreliable hearsay, obtained mostly five or more decades after the event, or are made without any evidence worthy of the name to support post-1905 regular collaboration by the couple on Einstein's researches.

THE JOFFE STATEMENT   In his discussion of the 1955 Joffe statement (see chapter 8), which he himself translates from the 1955 Soviet publication, Krstić does not explicitly state that Joffe actually saw the original manuscripts of the three groundbreaking papers mentioned in his memorial article. But the fact that the Russian physicist used the spelling "Marity" leads Krstić to write:

> My opinion is, that the academician Dr. A. F. Joffe used this special opportunity to tell to future generations something important with a deeper, lurking intention about Mileva's working relationship with Albert. Otherwise the question arises, how did he at all know of the family name Marity. The original manuscripts in Albert's handwriting of his important early articles have not been saved. It is quite possible that some parts among them were in Mileva's handwriting. (Krstić 2004, 114)

As noted earlier, the name "Marity" appeared before the publication of Joffe's memorial, in Carl Seelig's documentary biography of Einstein in 1954 (Seelig 1954, 29; 1956a, 24), and in any case Joffe could have come across the name in other circumstances (Stachel 2005, lx-lxi). The rest is self-evidently pure speculation and an example of the confirmation bias to be found throughout Krstić's book.

ORPHANED CLAIMS   Krstić asserts that Marić collaborated
with Einstein throughout their marriage, but he simply states a
number of those claims, including the three that follow, without
any sources at all.

[Just before the birth of Hans Albert in May 1904] they devoted
their evenings almost exclusively to working together on their scientific research. The result of such joint work was the article about
general molecular theory of heat [published in 1904]. (Krstić
2004, 101)

In this apartment [at Kramgasse 49 in Bern], their first son,
Hans Albert was born in 1904. One year later Mileva and Albert
formulated the basis of quantum theory with the definition of light
quanta. (Krstić 2004, 101)

While Albert was at the Federal Patent Office, Mileva took care
of the baby and the house, and she worked on science at home. After Hans Albert's birth the couple worked together even more intensively than before, but nearly exclusively in the evenings and at
night ... The result of such dedication was that in 1905 Albert sent
five [actually four] fundamental papers to the physics journal, Annalen der Physik. (Krstić 2004, 105–116)

Assertions made with such apparent authority from an author
"who studied Mileva's life and work with Einstein for some fifty
years" (Milentijević 2015, 123) may persuade the majority of
his readers, evidently including Milentijević (2015, 124, 134-
135, 143, 149), but for the more discerning among them they
indicate that Krstic's similar contentions throughout his book
should be treated with caution.

THE BROTHER'S REPORT   Krstić does make one apparently
substantive claim, with cited sources. It concerns a visit by
Marić's student brother Miloš with the Einsteins in Bern in early
1905 (Krstić 2004, 104-105, 216). That the visit definitely took

place is substantiated by Krstić's reference to a letter written by Miloš from the Einsteins' address, dated January 30, 1905, to a Professor Ostojić in Novi Sad. (He reports that the letter is now in a museum in Novi Sad.) Regarding the visit Krstić writes: "When he returned to Novi Sad Miloš told his parents and friends his observations of the time he spent in Bern. He described how the evenings and at night, when silence fell upon the town, the young married couple would sit together at the table, and by the light of a kerosene lantern, they would work together on physics problems. Miloš Jr. spoke of how they calculated, wrote, read and debated." At this point Krstić appends a footnote in which he writes: "Sidonja Gajin told me that in May of 1955. I was told the same thing by Sofija Golubović in July of 1961."

It is instructive to compare this account with Krstić's later brief discussion of what transpired from Miloš's visit::

> Mileva's brother, Miloš, had the opportunity to be "close by to observe" how Mileva and Albert lived and researched together. Miloš communicated his observations to his parents, who were always interested in and supported their children. It was already demonstrated [in "Appendix A," 2004, 214-220], that one could have confidence in Miloš's statements. The important echo of his narration reached me personally through talks with Mileva's sponsor, Sidonija Gajin, "kuma Sada," (1881–1973) [footnote 465], Mileva's cousin, Sofija Golubović, maiden name Galić (1889–1982), and Sofija's brother, Tima Galić (1902–1980) [footnote 466]. Miloš stated that Mileva and Albert would work regularly on physics problems in the evening and sit at the same table; they would calculate, write, read together and debate. (Krstić 2004, 216–217)

Note 465 reads: "I received Sidonija Gajin's statement soon after Albert Einstein's death, in May 1955." In note 466 Krstić writes that he "talked with Sofija Golubović in Belgrade ... in

July 1961, and also with Tima Galić in Novi Sad … in July 1961. Sofija was orphaned and lived with the Marić family, as her aunt was Marija Marić."

In the above passage Krstić is alluding to possibly one direct witness statement (from Golubović, who was living with Marić's parents) and two probably hearsay reports, with all the unreliability of such recollections from interested parties obtained more than fifty years after the event by an interviewer born in Novi Sad (Krstić 2004, 5) and so naturally predisposed to be biased in favor of the notion that Marić played an important role in Einstein's scientific achievements. In addition, we have no idea of how much of Krstić's initial, more detailed, account came from his interviewees and how much was his embellishment of what they said to him. It is also important to note that a close reading of Krstić's second report reveals that in regard to the quotation he gives he is not reporting what Miloš actually said but only an "important echo of his narration," which "reached [him]" from talks he had had with the three identified persons. Immediately following this he states that "Miloš stated …"—a classic example of the author leaving the reader with the impression that the quoted sentence comes from the just-mentioned interviewees without his explicitly stating that they actually reported those words to him.

In relation to these purported reports, an illustration of how accounts repeated by a second author may introduce elements not present in the original occurs in a paragraph by Milentijević in which she recycles Krstić's report of the 1905 visit: "Miloš had the opportunity to observe life in the Einstein household and wrote [sic] about it to his parents." On the basis of Krstić's initial account of the visit (2004, 104-105), Milentijević continues:

> When he returned to Novi Sad, Miloš talked to his family, relatives, and friends about his observations, reporting that the Einsteins

worked very hard and that Mileva especially appeared overburdened. Aside from household responsibilities and caring for the child, she worked with Albert on physics projects. Miloš described evenings at the Einstein home. After Hans Albert had been put to bed, Mileva and Albert would sit at a large table by the light of a kerosene lamp. They labored intensely on physics projects relating to relative motion, light quanta, and other topics. They read, discussed, calculated, wrote. Albert generally went to bed first because he had to get up early for work the next day, while Mileva stayed up late into the night doing mathematical calculations. Miloš' descriptions of the intense collaborative efforts of Mileva provided the basis for accounts among the relatives and friends in Novi Sad, Kać, and Titel about Mileva's contributions to Einstein's remarkable productivity. [Ref. Krstić 2004, 104–105] The result of such productivity was that in 1905 Albert published five [sic] papers in the *Annalen der Physik* that led to a scientific revolution and the onset of Einstein's fame. (Milentijević 2015, 115 and 442, n. 253)

The amount of embellishment here of Krstić's report from which Milentijević derives her considerably lengthier account is quite extraordinary, coming as it does in a book that is genuinely scholarly—except when the author deals with the years before and during the time when Marić and Einstein lived together. In fact, much of the above paragraph seems less an authentic account than an extract from a fictionalized story of the couple's life together. Unfortunately, even more than is the case with Krstić's book, the scholarly nature of most of Milentijević's book, and the numerous citations in the far less scholarly section up to the separation of Einstein and Marić in 1914, means that the reader is almost certainly going to take the above account to be evidence-based historical fact.

Returning to Krstić's second report of Miloš's visit to Bern in 1905, one of the interviewees he cites is Marić's cousin Tim Galić. It is of interest to note that Trbuhović-Gjurić reports that this same Galić recalled to her the baptism in Novi Sad of the Einsteins' two sons in September 1913 (T-G 1983, 106; 1988, 124). But she makes no mention of his having recollected what Miloš supposedly related to his parents in 1905. She would certainly have reported this in her book had he done so, especially since, only a few years earlier, Krstić states that *he* had obtained from Galić "an echo" of what Miloš Marić reported to his parents upon his return to Novi Sad from Bern in 1905 (Krstić 2004, 216–217). Trbuhović-Gjurić had obtained information from members of the Marić family and their friends when she was researching for her book in the 1960s (Trbuhović-Gjurić 1983 and 1988, 6), and she would surely have asked Marić's cousins if they had any memories relating to the Einstein couple. Undoubtedly she would also have spoken with those (including Sofija Golubović) who reportedly told Krstić about the Einstein' activities during the week-long summer holiday in Novi Sad and Kać in 1905 when they allegedly "would debate, work on mathematical computations, and write together" (Krstić 2004, 127). But nowhere in her book does Trbuhović-Gjurić provide accounts of either of these reported episodes, leading one to question the authenticity of the statements Krstić obtained from his interviewees.

In relation to what Miloš reportedly narrated after returning from Bern in 1905, Krstić urges his readers to "have confidence in Miloš's statements" since he was a "person whose statements one could fully trust," citing in support of this commendation the biographical sketch of Miloš in his "Appendix A" (Krstić 2004, 216, 105, 214–220). But, of course, it is not Miloš's trustworthiness that is at issue, it is the reliability of Krstić's interviewees' reported recollections of what Miloš narrated, obtained by Krstić some five decades after the events in question. It should now be evident that the story of what Miloš reported after returning from his visit with the Einsteins in 1905 is no more reliable than the more closely detailed (but false) dowry story related in chapter 8.

THE COLLABORATION INTENSIFIES   Krstić further writes:

> For the year 1906 there isn't any solid proof that Mileva and Albert collaborated in writing scientific articles, but according to the evidence of quoted eyewitness accounts about their joint work in 1905 and 1907, there is no doubt that their scientific collaboration had several years of continuity. On the basis of the previously mentioned letter Einstein wrote to Solovine dated 27th of April 1906, it can be concluded that [in 1906] the collaboration intensified. (Krstić 2004, 129)

In the letter Krstić cites, Einstein wrote: "My papers are much appreciated and are giving rise to further investigations. ... As for my social life, I have not been meeting with anyone since you left. Even the conversations with Besso have now come to an end" (CPAE 5, doc. 36). Although Einstein did not add that fortunately he still has his wife with whom he could discuss physics, Krstić (2004, 129) claims that "[h]e went straight home from the office to work jointly with Mileva in the evening and even during the nights," an assertion for which he provides no evidence.

WORKING AT THE SAME TABLE   In the Preface to his 2004 book, Krstić (2004, 16) writes: "Mileva worked regularly and during the nights, at the same table with Albert—quietly, modestly, and never in public view. The couple would discuss, calculate, and write together." Elsewhere, as we have seen, Krstić (2004, 204) refers to "several eyewitnesses," each of whom, he claims, has described in similar ways "the Einstein couple's manner of creating, sitting side by at the same table, absorbed in solving scientific problems." Leaving aside that (other than the suspect accounts of Marić's brother Miloš's narration in 1905) the only reports that he cites for this scenario were those of dubious evidential value he obtained from two people in relation to a *single* week the couple spent in Novi Sad and Kać (see chapter 8), and another equally suspect recollection he obtained from Hans Albert Einstein (see immediately below); Marić's letters to Kaufler Savić in the relevant period give no hint of any such scientific collaboration. For instance, in a letter to Kaufler Savić several months after their marriage, she wrote of her household chores and other activities (M-KS, 82–84), but there is no hint in this letter that she was working with Einstein on his scientific projects. Nor is there when she gave her friend brief reports of the papers he had published and the scientific work he was doing (M-KS 88, 94, 92, 100, 101)

The table-collaboration scenario also comes up in Krstić's account of two interviews he had in July 1971 with Einstein's eldest son Hans Albert. According to Krstić, the latter told him that his parents' "scientific collaboration continued into their marriage, and that he remembered seeing his parents work together in the evenings at the same table" (Krstić 2004, 7). However, only four years earlier, in a BBC radio program broadcast in 1967, when asked about his mother Hans Albert said, as recorded by the science historian G. J. Whitrow (1967, 19): "She was proud of him, but that was as far as it went. It was very hard to understand, because she originally had studied with him and had been a scientist herself. But, somehow or other, with the marriage she gave up practically all of her ambitions in that direction."

Hans Albert was born in 1904, and was five years old when, in 1909, Marić told Kaufler Savić of her unhappiness arising

Figure 9.1
Marić and Einstein in Prague, 1912.
Courtesy of ETH-Bibliothek Zürich, Bildarchiv (photo archive),
photographer unknown, Portr_03106. Public Domain Mark.

from Einstein's neglect of her in favor of his deep immersion in
his physics researches (M-KS, 94, 102). So neither of Hans
Albert's statements about his mother in the years following her
marriage could be reliable first-hand reports. But perhaps most
significant from our point of view is that Krstić obtained Hans
Albert's recollections only four years after the latter had given a
very different account on the BBC radio program.

A NECESSARY PARTNER

Krstić (2004, e.g., 61, 129, 204) claims that Marić collaborated
with Einstein on papers from 1900 to 1911 and beyond.[7] For
instance he writes that "for the year 1910 there is proof about

their joint work in Zurich" (Krstić 2004, 137), presumably referring to the Planck draft letter in Marić's handwriting. Likewise, Trbuhović-Gjurić writes that up to the time the couple left
Zurich for Prague in March 1911 (Seelig 1956a, 119), "she was
an absolutely necessary collaborator for him" (T-G 1983, 80;
1988, 98). But these contentions do not square with Marić's reporting to Kaufler Savić in summer 1909 that Einstein's new
post as professor of theoretical physics at the University of Zurich meant that he would be free of the eight hours daily toil at
the Bern patent office, and "will now be able to devote himself
to his beloved science, and *only* science" (M-KS. 94, emphasis in
the original). Nor do they accord with Marić's complaining to
her friend in winter 1909–10 that, with the increasing activities
arising from Einstein's growing fame among physicists, "he does
not have much time left for his wife" (M-KS, 100).

Finally, Krstić (2004, 129) asserts: "The production of papers
[in 1905 up to April 1906] exceeded the capacity of a single
man in a full-time job, which had nothing to do with physics."
It is true that Einstein's completion of four major papers, two of
which can be described as epoch making, in the span of little
more than three months between March 17 and June 1905 was
indeed astonishing. But Krstić fails to take into account that
many of Einstein's theoretical results contained in these papers
had been incubating for many years, in some cases extending
back to his student days. Moreover, his work at the patent office
was not particularly onerous, allowing him time to work surreptitiously on his own ideas during office hours (Fölsing 1997,
102; Isaacson 2007, 78).

In this chapter I continue the analysis of misleading, erroneous, or invented materials that have shaped the Mileva Story in three works: (1) a chapter (and relevant parts of an Introduction) about Marić in a book about women whose scientific and artistic accomplishments existed in the shadow of men; (2) a 2015 biography of Marić by a Serbian-born academic; and (3) a 2017 television docudrama that aired on the National Geographic network.

CHARLES S. CHIU, *WOMEN IN THE SHADOWS* (BOOK, 2008)

The propensity among writers on this subject to repeat statements made by their predecessors without attempting to confirm their accuracy, or reliability, is evident in "Mileva Einstein-Marić," the first chapter in Charles Chiu's *Women in the Shadows*, translated from the German by Edith Borchardt. Chiu manages to turn elements in the academic relationship between Einstein and Marić on their head, starting with the early period following their enrollment in the Zurich Polytechnic's teacher training program in 1896.

THE RETURN OF THE DARK-HAIRED WOMAN
As we saw in chapter 8, Aylesa Forsee created an imaginative scenario in the fictionalized account of Einstein's life she wrote

for young readers in which she described the first encounters between Einstein and Marić, "a young, black-haired Serbian woman." Trbuhović-Gjurić copied several of these imagined encounters, including invented dialogue, in her biography and added some imaginative misinformation of her own. Early in his chapter Chiu purports to describe the first contacts between the two students as follows:

> Einstein feels an immediate attraction to the dark-haired girl; her shyly reserved personality strongly appeals to him. But above all, he is impressed by her exceptional mathematical talent. When Mileva manages to get the result of an experiment which no one else in the group obtained, he—the youngest among the participants—addresses her, who is four years older: if she could help him with the works of Helmholtz, Maxwell, Boltzmann and Hertz? Not without hesitation, Mileva consents, and so they meet from then on for homework and discussions about God and the world. (Chiu 2008, 32)

As we have already seen, contrary to what Chiu writes Marić's pre-Polytechnic mathematics record was unexceptional, whereas Einstein was precociously gifted in the subject. The rest of the paragraph is pure fantasy, based, with his own further embellishment, on a highly imaginative passage in Trbuhović-Gjurić's book (T-G 1983 and 1988, 40–41).

MATHEMATICAL PROWESS
Continuing on the theme of mathematics, Chiu writes:

> Openly admitting his insufficiency in mathematics, [Einstein] now has found someone who is knowledgeable in exactly this discipline and can assist him in executing the mathematical aspects of his studies. And she knows her field, whether it is infinitesimal calculus, function theory, or force calculations. (Chiu 2008, 32)

Despite this supposed "insufficiency" in mathematics, Einstein had obtained maximum grades in the mathematical topics in the exams at the end of his final year at the secondary school at Aarau and in the Matura exams prior to entering Zurich Polytechnic. Equally fanciful is Chiu's depiction of Marić's mathematical accomplishments. Her limitations in this subject are apparent from her first-year semester grades in mathematical topics at Zurich Polytechnic, which can be found in the 1988 edition of Trbuhović-Gjurić's biography that Chiu lists among his "Works Cited." In that same book is a facsimile of Marić's Polytechnic Leaving Certificate grades, which also includes her rather mediocre 4.25 average grade for the mathematical components of the entrance exams that she took before entering the Polytechnic in 1896. Listed among the subjects on the later Leaving Certificate is the topic Differential and Integral Calculus, for which her grade was only 4.5. Ironically, in the case of another of the mathematical topics at which Chiu claims Marić excelled, Function Theory, she achieved a very poor 2.5 in her final diploma exams, which grade largely contributed to her failing to obtain a diploma in 1900 (Trbuhović-Gjuric 1988, 43; Zimmermann 1988a, 60–61, 64).

Later in the chapter Chiu (2008, 36) goes back to Marić's school career, claiming that "[a]s a sixteen-year-old, Mileva is brilliant in every subject—even in the Greek exam, feared by all, she passes with 'superior' ranking." The reference to the Greek exam identifies the school as the Royal Upper High School in Zagreb. The semester grades which Marić achieved there show that she was far from being outstanding in every subject, gaining only the equivalent of grade C (scale A–E) in most subjects in her last year—though Chiu does get one thing right: her only "excellent" grade in the two years she was at the school was in Greek (appendix A).

Following on from this, Chiu again enters the world of imaginative invention:

> At the beginning of 1894, Mileva Marić's life takes a new turn with many consequences. At eighteen she receives special permission from the school to participate in physics classes together with other regular—and that means male—students. Suddenly a new world is open to her, and this world fascinates her. Mechanics, optics, gravitation, and dynamics are her great passions from then on. The only girl in the seventh grade of the Superior Royal Academy at Agram [German name for Zagreb], she passes her final exam in mathematics and physics that very year at the top of her class. (Chiu 2008, 36)

Leaving aside Chiu's fantasies about Marić's "great passions" at that time, her final semester grades for mathematics and physics in the year she left the school were the equivalent of grade B (appendix A). In answer to a query from me, Mihaela Barbaric of the Zagreb State Archives wrote (October 15, 2013) that they have no record of the class placing of seventh grade students in the year in question.

Moving on to the Einstein couple's early years of marriage, Chiu (2008, 42) writes that while Einstein is evaluating patents at work, when time permits "[Marić] works on mathematical calculations requested by Albert, which he as a theoretician finds burdensome." This apparently is because, as he informs the reader a little later (Chiu 2008, 43), "Einstein ... always reluctantly ventured into the lowlands of differentials, integrals, and infinitesimals." This absurd statement is about someone who had mastered the basics of differential and integral calculus by the time he was fifteen (Talmey 1932, 163–164; Isaacson 2007, 19–20).

THE 1990 AAAS MEETING

In an epilogue to his chapter on Marić, Chiu turns to the session of the 1990 meeting of the AAAS in New Orleans during which the claims about Marić's alleged role in Einstein's work were presented and debated. According to the list of session speakers published in *Science*, the AAAS journal, they included: Robert S. Cohen, Caroline L. Herzenberg, Ruth H. Howes, Lewis R. Pyenson, John J. Stachel, Senta Troemel-Ploetz, and Evan Harris Walker (Herschman 1989, 1326). Chiu (2008, 53) alludes to "the newest research results of the three presenters, who claim that Einstein's first wife, indeed, contributed either 'only' mathematically or possibly even conceptually to the development of the 'theory of relativity.'" Two of his named "presenters" were Troemel-Ploetz and Walker. The third, he says, was the physicist and physics historian Abraham Pais, via a conference call from New York. However, Pais was not listed among the seven participants in the official program of the session, and there is no record that he called in to join the meeting (Herschman 1989, 1326). Had he done so, he certainly would have not supported the contentions of Troemel-Ploetz and Walker. Pais had previously seen a draft of the German edition of Trbuhović-Gjurić's book, and later expressed his view of it: "What is written about the role of Mileva in regard to Albert's scientific output is surprising and astonishing. I have not found a single reason for believing that in this respect the author's allegations are founded on fact" (Pais 1994, 2).

EDITH BORCHARDT, "INTRODUCTION" TO CHIU (2008)

The translator of Chiu's book, Edith Borchardt, a professor of German at the University of Minnesota, wrote an introduction to the book in which she prefigured several of Chiu's contentions, including the Pais reference, while adding some of her own. For instance, in regard to the 1905 special relativity paper, Borchardt (2008, 4) writes: "Trbuhović-Gjurić is convinced

that Mileva did not merely inspire Einstein but that she produced the mathematical proof for his theory." She is here recycling an assertion made by Trbuhović-Gjurić for which the latter
provides no evidence, though a few pages later she quotes the
unreferenced and unsubstantiated statement made by Peter Michelmore on similar lines in his novelistic book *Einstein: Profile
of the Man* (T-G 1983, 69, 72; 1988, 87, 90; Michelmore 1963,
41–42). As a noted journalist resorting to the novelistic style,
Michelmore apparently felt no necessity to provide evidence for
his manifestly invented words and scenarios; no doubt he would
have been horrified by the remarkable number of authors who
have treated these statements as historical fact.

In similar vein, a short time later Borchardt enlists Andrea
Gabor for a similar role. After quoting John Stachel's views on
the alleged contributions made by Marić to Einstein's scientific
work, she writes:

> [Gabor's] assessment of Mileva Marić differs from Stachel's views,
> since she believes with Einstein's biographer Peter Michelmore that
> Mileva was "'as good at mathematics as Marcel [Grossman
> (*sic*)].'"(Borchardt 2008, 6)

This reads as if Gabor had independently come to the same
conclusion as Michelmore, whereas she was merely quoting him
via Trbuhović-Gjurić (Gabor 1995, 25, 295). What we have
here is an author (Borchardt) quoting another author (Gabor),
who in turn was quoting a third author (Trbuhović-Gjurić),
who again in turn was quoting a fourth author (Michelmore),
whose assertion in question is demonstrably erroneous.

Borchardt's propensity to invoke hearsay claims regardless of
their provenance is exemplified by her quoting the words of one
Milenko Damjanov, who was an elderly relative of the daughter
of Marić's friend Milana Bota Stefanović, interviewed by

Michele Zackheim in the late 1990s: "Mileva was better in mathematics than her husband" (Borchardt 2008, 5; Zackheim 2000, 203). Borchardt follows up this item of non-evidence with another from an equally unreliable hearsay source: "This sentiment is echoed by Dr. Ljubomir-Bata Dumić, who wrote that Mileva solved all mathematical problems for Einstein, especially in regard to his theory of relativity" (Borchhardt 2008, 5; T-G 1983, 75; 1988, 93; see chapter 6).

In a passage earlier in this section Borchardt (2008, 2–3) outdoes these examples for sheer irrelevancy in what is a purportedly scholarly resumé of the evidence in support of the "Mileva Story." She provides excerpts from a 1999 play with the title "Mileva Einstein" by Vida Ognjenović which professes, as described by Borchardt, to evoke significant events in Marić's life including the "personal and professional struggles she had to face to pursue a career as a scientist." Borchardt continues: "Among these are convincing Dr. Weber, professor of physics at the Zurich Polytechnic, to let her attend his classes, since she passed the entrance examination and intended to be a student in his course." This is a bad start. Marić did not pass the entrance exams (she didn't need to take them as she had passed her Matura exams—see chapter 9), and having enrolled for the course for intending teachers of physics and mathematics, she attended Weber's (second year) physics course automatically. There follows (in the play) a scene in which, as described by Borchardt, Weber "makes clear to [Mileva] that 'the university is not a finishing school for young ladies' … [H]e continues his tirade, citing the laws of nature, 'which has created female beings in such a way that they cannot grasp the gist of abstract science.'" However, "Mileva insists on enrolling at the university in Zurich, since equality for men and women … was a right granted in Switzerland thirty years earlier." After more ranting from Weber, during which he "warns her that obtaining a higher degree 'would be too heavy a burden for the weak shoulders of a woman,'" he "finally concedes in the presence of the male students gathered around him (Albert Einstein, Marcel Grossman [*sic*], and Jakob Ehrat): 'Well, gentlemen, then let it be known! This young lady here is the pedagogical transgression of my life! This is the first female student of physics in the history of this serious and conservative school!'"

What is concerning is not just that this historical nonsense is presented to the reader as if it constituted evidence of the barriers Marić had to overcome to obtain higher education even in Switzerland. It also illustrates, to an extreme degree, the kind of gross misrepresentation of the true facts about Marić's life, especially in relation to Einstein's scientific achievements, that have become widely disseminated in recent times.

RADMILA MILENTIJEVIĆ, *MILEVA MARIĆ EINSTEIN* (BOOK, 2015)

Radmila Milentijević is a Serbian-born former professor of Modern European History at the City University of New York and has for more than two decades been the president of the World Serbian Voluntary Fund (Milentijević 2015, 430–431). Her 2015 *Mileva Marić Einstein: Life with Albert Einstein*—an impressive 489-page biography, with 1,350 endnote citations—is extremely well written and clearly involved a massive amount of research. She consults original sources in many instances and provides much valuable information, especially for the long period after Marić's separation and divorce from Einstein (Milentijević 2015, 174–414). Unfortunately, the early part of Milentijević's book (2015, 24–173), that dealing with the period before and during the years in which Marić and Einstein

lived together, is marred by numerous factual errors and misconceptions, only a small number of which can be dealt with
here. One of the deficiencies of this part of the book is that the
author has a propensity to quote uncritically from Michelmore,
Trbuhović-Gjurić, Walker, and Krstić as if their publications
were a reliable source of information, which is by no means the
case. She also occasionally makes assertions that have no evidential foundation, presented as if they were incontrovertible historical facts. I shall start with two less important instances.

MINOR ERRORS

Milentijević writes that Einstein and Marić "read books of their
own choosing in physics, a field in which they both performed
brilliantly on the final oral examination" (Milentijević 2015,
58). In fact, while their final diploma grades in experimental
physics were moderately good (both achieved 5 on a scale 1–6),
in theoretical physics they were 5 and 4.5 respectively (*CPAE* 1,
doc. 67). Far from being "brilliant," Marić's 4.5 in theoretical
physics can only be described as mediocre.

The next item concerns the Olympia Academy inaugurated
by Einstein, Maurice Solovine, and Conrad Habicht soon after
Einstein's arrival in Bern at the end of January 1902. They met
originally after Einstein had placed an advertisement in a local
newspaper offering private tutoring in mathematics and physics
while he waited for a hoped-for position at the Bern patent
office to be advertised. After a few lessons, Einstein and the
other two found each other's company so congenial that they
gave them up and began to meet regularly for a frugal meal and
the joint study of books on science and philosophy by eminent
authors, such as Ernst Mach, John Stuart Mill, David Hume,
and Henri Poincaré (Seelig 1956a, 57). After Marić married
Einstein in January 1903 she attended these so-called Olympia

Academy gatherings when they were held (as was usual) at the Einsteins' home. Solovine later wrote of the marriage: "This event caused no change in our meetings. Mileva, intelligent and reserved, listened attentively to us, but never intervened in our discussions" (Solovine 1956, xii).

According to Milentijević (2015, 112), "Mileva respectfully did not intrude in [the Academy] discussions because they were official sessions for which Solovine and Habicht each paid Einstein." However, she provides no citation for her assertion that Solovine and Habicht went on paying Einstein after the short period of their receiving lessons from him, and it has no factual basis (Fölsing 1997, 98–99; Isaacson 2007, 79–84).

MAJOR ERRORS

One of Milentijević's more significant errors is in regard to the Joffe Story. After quoting extensively from Evan Harris Walker's flawed arguments in his 1991 letter to *Physics Today* purportedly demonstrating that Marić was co-author of Einstein's three major 1905 papers on the basis of Joffe's memorial article to Einstein (see chapter 8), Milentijević (2015, 123) writes in support of his claims: "Joffe had worked in 1905 as an assistant to Wilhelm Roentgen, who was then on the board of the *Annalen der Physik*, and in this position had reviewed manuscripts before publication" (Milentijević 2015, 123). Although she does not cite Trbuhović-Gjurić, she evidently obtained this information from the latter's biography (T-G 1983, 79; 1988, 97). However, as demonstrated in the analysis of this story in chapter 8, the statement has no factual basis, and Milentijević makes no attempt to show that it does. She goes on to quote the key statement in Joffe's 1955 memorial article for Einstein, evidently in her own translation from the Russian:

"In 1905, the *Annalen der Physik* published three articles that laid down three fundamental principles for the development of physics in the twentieth century. They were: the theory of Brownian motion, the photoelectric effect and the special theory of relativity. Their author was until then an unknown clerk in the Patent Office in Bern, Einstein-Marity." In parentheses, Joffe explained that, "Marity was his wife's family name, which according to Swiss custom is added to the husband's family name." (Milentijević 2015, 123)

At this point one might have expected Milentijević to explain how what Joffe wrote justifies her writing that he had seen the three submitted papers when he was an assistant to Joffe, and why Roentgen, an experimentalist, would have been asked to review the manuscripts when the editor of the *Annalen der Physik* was the theorical physicist Paul Drude, and his advisor on theoretical physics papers was Max Planck (Stachel 2005, lix–lx). Instead she asserts: "There can be little doubt [*sic*] that Joffe saw the name Einstein-Marity inscribed on those papers, although he was misinformed that 'Marity' had been added to Albert's last name because of a Swiss custom."

What we have here is another example of an emphatic statement, without any evidence to justify it, made by the author of a scholarly work, that (together with her uncritical recycling at length of Walker's dubious contentions [Milentijević 2015, 122–123; see Stachel 2002, 26–29, 32–37]) will have undoubtedly convinced the great majority of her readers. The latter, of course, would be highly unlikely to have any knowledge of the scholarly rebuttals of Trbuhović-Gjurić's original claims (e.g., Stachel 2005, liv–lxiii; Martínez 2005, 51–52; 2011, 198–200), which are implicitly endorsed by Milentijević in the passage in question.

CLAIMS BY MICHELMORE AND TRBUHOVIĆ-GJURIĆ

After observing that the output of scientific ideas by Einstein in 1905 was "extraordinary," Milentijević (2015, 121) asks: "How was it possible for a full-time bureaucrat, who worked eight hours a day, six days a week at the Patent Office, to accomplish so much?" This has already been answered in chapter 9. Milentijević fails to take into account that for the most part the ideas in the 1905 papers had been several years in gestation— seven in the case of the relativity paper—and that at the patent office he was not only able to work on his ideas surreptitiously, he had three knowledgeable colleagues with whom he was able to discuss them (Fölsing 1997, 110–111, 115–116). However, Milentijević responds to the question by writing that "Peter Michelmore was the first of Einstein's biographers to acknowledge that Mileva's contribution was a factor in making it possible, and that she played a significant role in Einstein's achievements of 1905." She then repeats the much quoted, though demonstrably false, statement by Michelmore: "[Mileva] was as good at mathematics as Marcel [Grossmann]," She follows this up by quoting the same author's assertion that Marić "helped [Einstein] solve certain mathematical problems" in relation to the relativity paper (Michelmore 1963, 41). Her reference to Michelmore in this context as one of Einstein's biographers gives the false impression that his statements should be taken seriously as the product of scholarly research, when, as we have seen, his "Profile" of Einstein is a novelistic work in which occur numerous imagined scenarios with invented dialogue (see chapter 5).

Milentijević (2015, 121) follows up the quotations from Michelmore by writing that Trbuhović-Gjurić "credited Mileva to a greater extent with directly contributing to Einstein's achievements," adding that "Đurić-Gjurić also pointed out that

'Albert's later work emanated from the achievements realized in the period of his direct collaboration with Mileva'"—as if Trbuhović-Gjurić's statement was based on documented evidence rather than its being an evidence-free assertion. Milentijević continues: "As to Mileva's role in the creative part of the scientific discoveries of the *annus mirabilis*, Đurić-Trbuhović wrote: 'In Albert's work Mileva did not participate as a creator, no one else could have, but she tested his every thought, discussed it and provided the mathematical expression of his ideas on the extension of Max Planck's quantum and the special theory of relativity.'" This statement of Trbuhović-Gjurić's is nothing more than imaginative wishful thinking, with no evidential basis.

Milentijević (2015, 121–122) justifies her quoting Michelmore by noting that he "had the benefit of an interview in 1962 with Hans Albert [Einstein]," ignoring the fact that much of the early part of his book comprises a fictionalized version of Einstein's life. As for Trbuhović-Gjurić, Milentijević writes that she "relied primarily on the testimony of the many people she interviewed for the book." We have seen how utterly unreliable is such testimony from interested parties, demonstrated by Krstić's purportedly obtaining witness and hearsay accounts of scenes of the couple working on physics together "at the same table" during their week's holiday visiting Marić's parents in the summer of 1905, while Trbuhović-Gjurić reports no such recollections from the very same people (see chapter 9).

THE DIVORCE DECREE AS ALLEGED EVIDENCE FOR THE
MILEVA STORY

Marić and Einstein were separated in 1914 and divorced in 1919. According to Milentijević:

> Mileva asserted her right to recognition [of her contributions to Albert's scientific achievements] in the Divorce Agreement incorpo-

rated in the Divorce Decree of February 14, 1919, which designated the Nobel Prize as Mileva's property. It was through hard
negotiations that she was able to gain this measure of recognition
for her contributions to Albert's scientific achievements.
(Milentijević 2015, 126)

To confirm the statement in the first sentence of the quotation above, Milentijević cites her Appendix, which contains an
English translation of the Divorce Decree in full (Milentijević
2015, 418–423). She is correct in saying that it stipulates in
clause 4 that the anticipated Nobel Prize money "shall become
the property of Mrs. Mileva Einstein."[1] But, as we saw in chapter 6, it was not quite that straightforward: clause 4 of the decree
stipulates that Prof. Einstein "shall deposit this capital in trust at
a Swiss bank," and clause 4a stipulates that "Mrs. Einstein shall
have no authority over the capital without the consent of Prof.
Einstein," though she "shall dispose freely of the interest." Moreover, clause 4b stipulates that in "the case of the remarriage or
death of Mrs. Einstein … the Nobel Prize ceded to Mrs. Einstein … shall go to the children, Hans Albert and Eduard."[2]
Nevertheless, since Einstein almost invariably acceded to Marić's
requests when she required large sums of money (Milentijević
2015, 273, 279, 281, 321, 326), Milentijević's statement in the
second clause of the first sentence is essentially accurate.

The second sentence in Milentijević's statement above—"It
was through hard negotiations that she [Marić] was able to gain
recognition for her contributions to Albert's scientific achievements"—is without foundation. Leaving aside that Marić could
not access the Nobel Prize capital without Einstein's approval, we
have noted earlier (chapter 6) that, in the abundant correspondence between three of Einsteins' mutual friends who were acting
as intermediaries in regard to the impending divorce provisions,

there was no mention of Marić's raising the issue of her supposed contributions to Einstein's scientific work as a bargaining point in the protracted negotiations (Medicus 1994, 470).

THE 1925 LETTERS

Milentijević (2015, 287) contends in relation to the divorce settlement that letters from Einstein to Marić in fall 1925 provide evidence that she had made "contributions to Albert's earlier scientific achievements."[3] I shall examine here relevant passages from Milentijević's translations of extensive extracts from these letters in order to be in a position to assess the validity of her contention.[4] She introduces the episode in question by writing that, out of the blue, Einstein sent a letter to Marić (dated September 26, 1925) telling her that he was "drafting his Last Will"[5] and requesting from her and their two sons "a notarized declaration that upon receipt of the Nobel Prize [money], they would renounce any further claims on his estate, 'except what I declare expressly [I am] leaving to my sons in my Will.'" (The *Collected Papers*, volume 5, has "after having received the Nobel Prize [money]" (*CPAE* 15, Engl., doc. 79) in place of "upon receipt of" —Marić had already used substantial amounts of it by this time [Milentijević 2015, 273, 279, 281].)

Milentijević writes, not entirely accurately (see above), that "[u]nder the terms of the Divorce Decree, the Nobel Prize money was Mileva's property, and she viewed this entitlement as a right she had earned." According to Milentijević, Einstein's request "led Mileva to consider the need for documenting her contributions to Albert's earlier scientific achievements, possibly with the intention of making her claims public." She continues: "Presumably with this in mind, she had asked Albert for copies of the articles published in the *Annalen der Physik* on which they had worked together and asked him to send the publications in question to Zurich with Hans Albert, who was visiting his father at the time."

Milentijević comments (2015, 287): "Although this letter from Mileva to Albert has not survived, Albert's carefully crafted response dated October 24, [1925] has been preserved. 'I would feel like a criminal, when I read a letter from you, if I did not remind myself of the actual circumstances,' Albert declared. Had she earlier given Hans Albert the assignment to take with him the desired publications, Albert would have given them to his son or mailed them to Mileva. Now, however, 'there remain no more reprints of precisely my most significant works.' Moreover, '[i]f I send you the few most important works of which I have reprints, that does you little good, and you wouldn't read them.' Albert then reminded Mileva of what he had done for her in the past and what he was still doing for her."

Parenthetically, it should be noted that Milentijević's translation of the first clause of Einstein's last sentence above is inaccurate. An accurate translation reads: "If I send you the less significant articles, of which I have offprints..." (*CPAE* 15, Engl., doc. 95; "Wenn ich Dir die weniger bedeutendsten Arbeiten senden, von denen ich Separata habe" (*CPAE* 15, doc. 95).

Continuing with Milentijević's narrative, following her quoting some of the things Einstein said he had done for Marić, she writes: "Albert then subjected Mileva to an onslaught, ruthless even by his standards." She then quotes the relevant excerpt from Einstein's October 24 letter:

> But you really gave me a good laugh when you threatened me with your memories. Doesn't it ever dawn upon you for even a single second that no one would pay the least attention at all to your rubbish if the man with whom you are dealing had not perchance accomplished something important? When a person is a nonentity,

there's nothing more to be said, but one should be modest and shut up. I advise you to do so. (Milentijević 2015, 288; *CPAE* 15, doc. 95)

After noting that Einstein followed this attack with more conciliatory words, Milentijević describes what she believes were defects in Einstein's character: "Albert was exceptionally arrogant, self-centered and inconsiderate. When provoked, he hit back viciously and without mercy." Nevertheless, she writes, "Albert in his own way cared about Mileva. He respected her mind, sought the comfort of her company, punished her severely when she challenged him, and tried to be of help when she needed it." She continues: "Mileva responded to Albert's cutting remarks in a manner so gentle and understanding that it had a calming effect on him. Writing to her on November 1, Albert again requested the binding declaration, assuring her that his request did not reflect any hostility to her or the boys. He went on to explain his previous outburst":

> What infuriated me in your last letter was the threat with memories. What I cannot stand is a wide-ranging discussion of personal matters. … I was pleased, however, that you didn't return to that but instead wrote very nicely and politely." (Milentijević 2015, 289, ellipses in original; *CPAE* 15, doc. 99)

We are now in a position to judge Milentijević's stating, prior to her providing relevant extracts from Einstein's letters, that his request for notarized declarations "led Mileva to consider documenting her contributions to Einstein's earlier scientific achievements," and that it was "[p]resumably with this in mind she had asked for copies of the articles published in the *Annalen der Physik* on which they had worked together." It is evident that the crucial elements in these sentences of Milentijević's, the references to Marić's collaboration with Einstein on his earlier

scientific achievements, are nothing more than tendentious surmises on her part. The way that she assimilates into the sentence
in question the notion that Marić had asked specifically for copies of articles on which she and Einstein had collaborated makes
it appear as a factual statement based on a valid interpretation of
Einstein's words. However, in the absence of Marić's letters in
reply to Einstein's, she has no way of knowing what was in
Marić's mind, or what copies of Einstein's papers she had
requested. (In any case, the mere possession of offprints of some
of Einstein's papers would prove nothing.)

From Einstein's letter of October 24, 1925, we learn that
Marić had indicated to him that she was thinking of publishing
her memoirs, which Einstein chose to view as a threat, though
one which he told her he faced with "serenity" (*CPAE* 15, Engl.,
doc. 95).[6] In his letter dated November 1, 1925, he told her that
his requesting signed notaries "was not a hostile attitude to you
and the boys. I merely would like to prevent my current wife
and her daughters from being reduced to poverty due to negli-
gence on my part." He goes on to say that what enraged him
about Marić's last letter "was the threat of memoirs," and
explains that he had been "really annoyed with the nonsense by
Moszkowski," a popular journalist who wrote a book in 1920
that included extensive portions of several interviews that he
had had with Einstein (Moszkowski 1921)[7] (*CPAE* 15, Engl.,
doc. 99) It is also possible that Einstein's angry response to
Marić's suggestion derived, not just from his dislike of public
discussion of personal matters, but from a concern that she
might well include details of his less admirable behavior, for
example, his unfeeling neglect of her in favor of his immersion
in his pioneering research, his occasional lack of empathy in
regard to his teenage son Eduard's incipient mental health problems, and his starting an affair (conducted almost entirely by

mail) with his cousin Elsa Löwenthal in 1912 while he was still married to Marić (Highfield and Carter 1993, 147–149, 151).

MORE PROBLEMATIC CLAIMS

In April 1908 Marić took Hans Albert to visit her parents over the Easter holidays (*CPAE* 5, doc. 96). According to Milentijević (2015, 134), "[Jakob] Laub, a graduate student, ... replaced Mileva during her absence. While Albert put in his eight hours at the Patent Office, Laub stayed at the Einstein home and did calculations, as Mileva had done through the years." A little later she writes: "When Mileva returned from Vojvodina, Laub was gone. ... After completing the daily chores of housekeeping and putting Hans Albert to bed, Mileva took up again the calculations for Albert's ongoing scientific projects that Laub had performed in her absence" (Milentijević 2015, 135). This may seem straightforward enough. The twenty-six-year-old Laub, who recently received his doctorate and was now an Assistant to Wilhelm Wien, did stay in the Einstein home for at least five weeks in April to May 1908;[8] he was working with Einstein on two papers that they jointly submitted for publication dated April 29 and May 7, 1908 (CPAE 2, docs. 51, 52). But Milentijević provides no evidence to justify her writing that Laub was gone when Marić returned from Novi Sad, and in fact there is evidence to the contrary. A letter from Laub to Einstein dated March 1, 1908, indicates he would be in Bern to stay with the Einsteins at the beginning of April 1908. Einstein wrote a letter to Marić dated Good Friday [April 17, 1908] the contents of which indicate that Laub had been staying at their home for some time, that Marić had been at her parents' place for several days, and that Einstein was expecting her back soon. However, Laub left Bern for his home in Würzburg in Germany around the middle of May. It is clear from these dates that Marić must

have returned to Bern some considerable time before Laub left
(*CPAE* 5, docs. 91, 96, 101; Fölsing 1997, 239, 240). Likewise,
Milentijević provides no evidence that Marić had done Einstein's calculations "through the years"; that Laub was doing any
work that she normally did; or that Marić "took up again the
calculations for Albert's ongoing scientific projects that Laub
had performed in her absence."

It is instructive to look more closely at Milentijević's claims
cited above. All but one of them refer to Marić's doing "calculations" for Einstein's scientific projects, by which she apparently
means arithmetic calculations—otherwise they would be indis-
tinguishable from Einstein's scientific work in general. These
claims evidently are based on two misconceptions. One is that
Marić was so proficient in mathematics that Einstein needed her
to undertake "calculations" in tandem with his work on theoretical physics. This notion is presumably based on Milentijević's
taking as historical fact the erroneous assertion by Michelmore,
recycled by Trbuhović-Gjurić, that Marić excelled in mathematics (Milentijević 2015, 121; Michelmore 1963, 31, 41;
Trbuhović-Gjurić 1983, 72, 75, 87; 1988, 90, 93, 105). The
other misconception is that Einstein's many-sided contributions
to topics on the frontiers of contemporary physics constantly
required arithmetical calculations to complement his theoretical
work. While this was occasionally the case (most notably when
he needed to use empirical data obtained by other physicists to
check the validity of his theoretical conclusions), most of the
time there were no arithmetical calculations involved (see, e.g.,
CPAE 5, docs. 153, 163, 178, 188). This misconception could
only be held by someone who, like most people (including the
great majority of academics), has little idea of the nature of Einstein's scientific research and of the essence of theoretical physics
in general.

Finally in this section, Milentijević offers a seemingly plausible (from her point of view) explanation for why the 1901 paper on capillarity (*Einstein 1901*, *CPAE 2*, doc. 1) was published under Einstein's name alone:

> A scientific paper resulting from a partnership in today's world would have been authored by both scientists or, at a minimum, each would have received credit for his or her contribution. The only reasonable conclusion is that although this publication was the result of their joint labor, Mileva and Albert had agreed that the article would be published under Albert's name alone. Why? Albert was without a job. His personality and earlier conduct at the Polytechnic was now seriously impeding his chances of obtaining employment. The only way to overcome this disadvantage was for Albert to demonstrate that he was a respected scientist and thereby establish his name in the academic community. To achieve this goal, he needed Mileva's help. ... Mileva and Albert were resolved to get married, and a precondition for marriage was that Albert secure a job. The couple had decided to join forces for Albert's success. Mileva was willing to dedicate herself totally, using her talents and her time to advance Albert, and giving Albert full credit for their joint efforts for the sake of their future together. That Albert never gave recognition for her contribution to his achievements is a reflection on his character. (Milentijević 2015, 69–70)

It is ironic that Milentijević wrote these words in relation to a paper for which, as I previously demonstrated (see chapter 7), no documented evidence exists to support the claim that Marić contributed to it. Moreover, it is in relation to this paper that Marić reported to her closest friend Helene Kaufler Savić in December 1990: "Albert wrote a paper in physics that will probably soon be published in the *Annalen der Physik*. You can imagine how proud I am of my darling" (M-KS, 70). These are not the words of someone who has collaborated in its production. More generally, as previously noted, Marić's letters to Kaufler Savić during the years during which she lived with Einstein are full of her current preoccupations, but she never so much as hints that these included assisting her husband in his researches on theoretical physics. On the contrary, when she mentions Einstein's published papers, or his scientific work, she attributes them unequivocally to him (M-KS, 70, 88, 101, 108). There is no reason not to take literally Marić's words to her friend in December 1906, when she wrote, after recounting the antics of the infant Hans Albert, "the papers [my husband] has written are already mounting quite high" (M-KS, 88). And in 1909, after Einstein had obtained his first university teaching post, she wrote with signs of resentment for his neglect of her, "he will now be able to devote himself to his beloved science, and *only* science" (M-KS, 94. emphasis in the original). Yet Milentijević would have her readers believe that Marić collaborated continuously with Einstein on his scientific research right up to the birth of their second son Eduard in July 1910 (Milentijević 2015, 112, 115, 117, 120, 121–126, 130, 134–135, 143–144, 149).

NATIONAL GEOGRAPHIC CHANNEL, *GENIUS: EINSTEIN* (TELEVISED DOCUDRAMA, 2017)

This ten-episode docudrama about Einstein, part of the National Geographic series *Genius*, is probably the most widely familiar portrayal today of Einstein's private life and public work and engagement (Howard et al., 2017). It is based on Walter Isaacson's (2007) generally reliable popular biography of Einstein. The drama's dual focus is on Einstein's early work and life with Marić and on his much later life as the famous genius; and the episodes sometimes jump disconcertingly between the two. But it stays focused long enough in the early episodes for the drama's writers to convey to their viewers their conceptions of Einstein's personal life and his relationship with Mileva Marić.

The first episode opens with two attention-grabbing scenes that, in the case of the second one, ensure that viewers are fully aware of the less admirable side of the life of the great man before they even encounter him. The dark forces of the first half of the twentieth century that so impacted the course of Einstein's life are embodied in the first scene by the brutal assassination in 1922 of Germany's Jewish foreign minister, Einstein's friend Walther Rathenau, by Nazi thugs, an event that greatly affected Einstein. The next scene, all but pandering to the debunking spirit of the age, bars interested young people from viewing (at least officially) by portraying the middle-aged Einstein in a TV 14–rated sex scene with his much younger secretary Betty Neumann which must have been situated in the latter part of 1923 (Fölsing 1997, 548). When Neumann refuses Einstein's absurd invitation to move in with him and his wife Elsa (*CPAE* 14, doc. 140), she points out, "You have a wife," to which he replies, "One whom I adore." "You're insane!" she shouts at the thought of a ménage à trois. Then she remarks, "For a man who is an expert on the universe, you don't know the first thing about people, do you?"

In the following scenes, Marić frequently appears in a nightgown and even in bed with Einstein, rather than studying for her exams, which contributed, the writers suggest, to her failing performance. There is no factual basis for any of this. Their love was not consummated until May 1901 (E-M, 51), nearly a year after Marić had failed the Polytechnic final diploma exams in July 1900, and they did not live together until their marriage in January 1903.

Episode two has a scene purporting to show what happened when Einstein met Marić at the beginning of their first semester at Zurich Polytechnic in October 1896. Before he had time to introduce himself, Marić launches into a brief recital of her knowledge of Greek cosmogony (the science of the origin of the universe), followed by a listing of the eminent physicists, Kepler, Newton, Faraday, and Maxwell. In another display of erudition, in regard to Maxwell she exhibits her knowledge of the mathematical form of his electromagnetic theory by reciting one of his famous equations. This scene has as much authenticity as Chiu's highly imaginative account of their first meeting (above). In fact, this docudrama version of their meeting for an adult audience is as fantastical as Forsee's invented account of Einstein and Marić's early acquaintanceship written for children and young adults (see chapter 8).

There follows a scene in which professor Weber introduces Einstein to Marić at their first class: "I see you haven't met the only student to score higher than you on the mathematics section of our entrance examinations." Evidently historical accuracy was not a priority for the writers of the early episodes of this film. As already reported (chapter 9), when Einstein took the entrance exams in 1895 he was only sixteen and had been out of the school system for some nine months. He failed on most subjects, but achieved excellent grades in mathematics and physics. Marić, having already passed the Matura exams in 1896, did not have to take the Polytechnic entrance exams, though she was required to take the mathematical subjects. On these, her average grade was a rather mediocre 4.25 on a scale 1–6. Again, the students in class VIA, the mathematics and physics section of the Polytechnic teaching diploma courses, did not start classes with the physics professor Heinrich Weber until their second year at the Polytechnic, but this is a minor error compared with the fabrications perpetrated in other scenes involving Marić and Einstein.

Finally, we come to episode 4, on the so-called miracle year of 1905. In one scene we observe the young Einstein becoming increasingly discouraged that his radical ideas about light quanta, the basis of one of his major papers of 1905, are completely rejected by his patent office colleagues and by his University of Zurich doctoral adviser, Alfred Kleiner. Dejected, he returns home to his Bern apartment where Marić, baby Hans Albert, and both of Marić's visiting parents are present. Einstein expresses his frustration when he asks: "Who would publish the work of a third-class patent clerk?"

None of this has any basis in historical facts. When he was at the patent office, as noted above, he had discussions on theoretical physics with three of his colleagues, Josef Sauter, Paul Gruner, and Michele Besso, all of whom were receptive to his ideas (Fölsing 1997, 110–111, 113, 123, 155). Einstein's doctoral dissertation was on the determination of molecular dimensions, and there is no evidence that he ever discussed light quanta with his doctoral advisor Alfred Kleiner. Again, Marić's parents did not visit the Einsteins' home in Bern in 1905, nor indeed, at any time. In regard to Einstein's *cri de coeur* question above, he had already had five papers published in the *Annalen der Physik* prior to 1905 (Fölsing 1997, 825).

As if this were not enough, there is a scene in which, after the publication of the 1905 relativity paper, Marić is depicted as being distressed because she helped with much of the research for the paper, but only Michele Besso was acknowledged by Einstein in the paper. The contention that Marić assisted Einstein in his research that led to his special relativity theory again has no evidential basis. In short, the screening of the early episodes of *Genius: Einstein* have had the unfortunate effect of disseminating the Mileva Story to a wide public, on this occasion under the imprimatur of the highly respected *National Geographic* magazine.

11 THE STORY CONCLUDES

In this three-part book David C. Cassidy and I have examined the story of Mileva Einstein-Marić from different perspectives. In part I, Cassidy introduced Marić and the story of her relationship with Einstein as revealed in a narrative he based on old and new documentary evidence. In part III, I provided a critical examination of the Mileva Story that has been disseminated widely over the past nearly three decades regarding her supposed scientific collaboration with Einstein before and during their marriage. This necessitated a thoroughgoing survey of relevant documents and the scholarly literature, together with the close reading of a variety of publications promoting the Mileva Story. Between these chapters, in part II, Ruth Lewin Sime contributed an insightful account of the broader struggle of women like Mileva Marić to enter scientific fields during the nineteenth century and first half of the twentieth century.

My critique undermines many of the claims that have been repeated in support of the Mileva Story. Cassidy's account constitutes, we believe, a more realistic and compelling portrait of this remarkable woman as she sought to overcome many obstacles in her determination to pursue a career in science. Tragically, as we have seen, for a variety of reasons she did not achieve her full potential as a scientist, or as a science teacher, nor did she realize her hopes and dreams in marriage and in life.

Hers is a story that is not often told. Yet, when closely examined, it provides important and valuable insights into the struggles of this determined woman then, and for some even nowadays, for status and recognition in science. On the other hand, it is ironic, as Gerald Holton (2000, 191) has pointed out, that the accounts exaggerating Marić's academic prowess and her purported role in Einstein's publications "only detract both from her real and significant place in history and from the tragic unfulfillment of her early promise. For she was one of the pioneers in the movement to bring women into science, even if she did not reap its benefits."

The various published accounts have thus far been problematic because most authors have taken too much on trust from the earliest book on the subject of Mileva Marić, a biography by the Serbian academic Desanka Trbuhović-Gjurić, published in Yugoslavia in 1969, and later issued in more widely accessible German and French language editions in the 1980s and early 1990s. In 1990 a lengthy article in English published by the linguist Senta Troemel-Ploetz summarizing the essence of Trbuhović-Gjurić's biography drew attention to her book. Whatever its shortcomings, Trbuhović-Gjurić's biography had the considerable merit of belatedly drawing attention to the existence of Einstein's first wife, who was previously given only a passing mention in biographies of the great physicist. Unfortunately, since the publication of Troemel-Ploetz's article, several book chapters about Marić have been published by authors who evidently made no effort to seek out reliable documentary material to check the numerous contentions made by Trbuhović-Gjurić and repeated by Troemel-Ploetz.

To gain an insight into Mileva Marić's place in history it is necessary to summarize her life story, as has been ably accomplished by Cassidy. Born at a time when women faced high hurdles for entering into higher education and careers in science, she had the added disadvantage of a congenital dislocated hip, which left her with a lifelong limp that was to be a cause of inner withdrawal during her early education. But, on the other hand, she had the advantage of a very supportive father who recognized her early gifts for language and simple calculations, and who encouraged her through to the end of her secondary schooling, where her academic achievements were generally good, though not outstanding. After that the most fateful event was her entry into the Zurich Polytechnic in 1896 at the same time as Albert Einstein. They both majored in physics and joined a small group studying for a teaching diploma (the other students in the class majored in mathematics). By the second year of their acquaintanceship, the letters exchanged between them during vacations reveal that she had struck up a burgeoning relationship with the scientifically ambitious young Einstein. During the second year of the four-year course, Einstein became dissatisfied with the material presented by the head physics professor, Heinrich Weber, as it failed to encompass up-to-date theoretical advances of which his extracurricular reading had led him to be aware. From then on, at Einstein's instigation, the couple studied books by eminent physicists together, and he enthusiastically communicated to Marić in his letters his ideas on a variety of physics topics. Unfortunately, evidence is lacking of any contributions that Marić may have made to these ideas at that time since Einstein did not save most of her letters. However, Einstein's own letters give no indication that in the discarded letters she had made any contributions to his theorizing. (As I have previously noted, in the two cases where we have her letters responding directly to Einstein's which contain passages

about his latest ideas on his extra-curricular researches, Marić
made no response to these and wrote only of personal matters
[E-M, 10–11, 12–13, 47–48, 48].)

Marić's career ambitions were set back by her failing the final
diploma examinations in 1900, largely due to her poor grade in
the mathematics component (theory of functions), plus her
mediocre, heavily weighted, grade for her diploma thesis. She
failed again when she retook the exams the following year, at a
time when she was some three months pregnant with Einstein's
child and suffering considerable emotional distress because of
the long periods when she was unable to see him. (He was either
with his parents in Italy, or, later on, elsewhere in Switzerland
where he had managed to obtain temporary teaching posts.) In
addition, her attempt to complete a doctoral dissertation, based
on extending her 1900 diploma thesis, was prematurely ended
following disagreements with her physics adviser, Heinrich
Weber. Einstein was eventually able to obtain a post at the Bern
Patent Office in June 1902. Marić, soon to lose her out-of-wedlock infant daughter (to scarlet fever or possibly adoption),
joined him in Bern in January 1903, in which month they were
married in a ceremony at which none of their respective parents
attended. Their married life began with Marić, although elated
at her long-delayed union, still carrying the mental scars of her
lost daughter and her academic failures. There soon followed the
birth of their first son, Hans Albert, a source of joy for Marić.
Although there is no hard evidence to this effect, it is very likely
that she would have undertaken essential tasks to assist Einstein
with his theoretical work, such as proofreading papers, checking
out physics texts, and so on. However, she made no mention of
any *scientific* assistance given to her husband when relating her
current preoccupations in letters to her closest friend Helene
Kaufler Savić. In addition to the practical help she likely gave

Einstein in relation to his scientific work, during these earlier contented years together she provided a loving and stable home life that was absolutely essential for Einstein if he was to be able to immerse himself in his researches in addition to his full-time employment (albeit not too arduous), six days a week, at the patent office. To what extent Einstein discussed his ideas with her in the early years of their marriage is impossible to ascertain from the available evidence. Be that as it may, she certainly maintained a close interest in Einstein's scientific achievements that continued up to their separation in 1914 and beyond. What can be said with certainty is that her role was invaluable in providing the circumstances that were essential to enable the coming to fruition of the ideas that flowed abundantly from Einstein. Together with this crucial contribution, her determination to obtain a scientific education, almost unique for young women in that era, merits for her a permanent place in the history of physics in the early years of the twentieth century.

APPENDIX A
MARIĆ'S PRE-POLYTECHNIC GRADES

MARIĆ'S SEMESTER GRADES AT THE ROYAL
CLASSICAL GYMNASIUM, ZAGREB
(TRANSLATION OF FIGURES 1.1 AND 1.2)

Grade scale: excellent (A), very good (B), good (C), satisfactory (D), unsatisfactory (F)

	1892–93		1893–94	
Semester	I.	II.	I.	II.
Exam date	26 Jan.	16 June	25 Jan.	4 Sept.
Catechism	Very good	Very good	Good	Good
Latin	Good	Good	Good	Good
Greek	Excellent	Very good	Good	Good
Croatian	Good	Good	Good	Good
German	Very good	Very good	Satisfactory	Good
History and Geography	Very good	Very good	Good	Good
Mathematics	Very good	Very good	Good	Very good
Biology and Zoology	Good	Good		
Physics			Satisfactory	Very good
Propaedeutics			Very good	Good

Source: Figures 1.1 and 1.2. Archival source in captions.

MARIĆ'S GRADES ON THE MATHEMATICS ENTRANCE
EXAMS, ZURICH POLYTECHNIC, FALL 1896

**The scale is from 1 to 6 (highest); multiple grades may indicate
multiple parts.**

Mathematics	5	3.5
	4.5	5
Descriptive Geometry	4	3.5

Source: Marić, Student Record; T-G 1983, 56; T-G 1988, 60

APPENDIX B
EINSTEIN'S PRE-POLYTECHNIC GRADES

EINSTEIN'S SEMESTER GRADES FOR SELECTED SUBJECTS
AT THE AARGAU CANTONAL SCHOOL

**For 1895–96 the grade scale was 6 (lowest) to 1 (highest). For
1896–97 the scale was reversed.**

	1896–97		1896–97
Semester	**III.**	**IV.**	**I.**
Arithmetic and Algebra Geometry	1	1	6
Descriptive Geometry	3	3–4, 2	4, 4–5
Physics	1–2	1–2	6–5
Chemistry	1, –	2, 3	5

Source: *CPAE* 1, doc. 10.

EINSTEIN'S FINAL GRADES FOR SELECTED SUBJECTS,
AARGAU CANTONAL SCHOOL, FALL 1896

The scale is from 1 (lowest) to 6 (highest)

Algebra	6
Geometry	6
Descriptive Geometry	5
Physics	5-6
Chemistry	5
French	3

Source: *CPAE* 1, doc. 19.

EINSTEIN'S GRADES ON THE WRITTEN PORTION
OF THE MATURA EXAMS

The scale is from 1 (lowest) to 6 (highest)

German	5
French	3-4
Geometry	6
Physics	5-6
Natural History	5
Algebra	6
Chemistry	5

Source: *CPAE* 1, docs. 21–27.

APPENDIX C
SEMESTER GRADES FOR EINSTEIN AND MARIĆ AT THE ZURICH POLYTECHNIC

The scale is from 1 (lowest) to 6 (highest). Ungraded courses are not included.

	Marić		Einstein		AE Remark
1896–97	I	II	I	II	
Differential & Integral Calculus w. Exercises	4.5	4.5	4.5	5	
Analytic Geometry	4.5		5		
Descriptive Geometry w. Exercises	4.5	4	4.5	4	
Projective Geometry		3.5		4.5	
Mechanics w. Exercises		4.5		5	
1897–98					
Differential Equations			5		
Weber, Physics		5	5.5	5	
Mechanics			5.5		
Projective Geometry			4		
1898–99					
Weber, Electrotechnical Laboratory			6		
Pernet, Physics Practicum for Beginners	5		1		March 1899: Director's reprimand for non-diligence

	Marić		Einstein		AE Remark
1896–97 *(cont.)*	I	II	I	II	
Weber, Scientific Projects in Physics Laboratories		5		5	
Determination of Geographical Location		5		4.5	
1899–1900					
Weber, Scientific Projects in Physics Laboratories	6	5	6	5	
Geometry of Position	5				
Summer 1901					
Determination of Geographical Location	5				
Weber, Scientific Projects in Physics Laboratories	5				

Sources: For Marić, Student Record; for Einstein, *CPAE*, doc. 28.

APPENDIX D
GRADES ON THE INTERMEDIATE
AND DIPLOMA EXAMS

GRADES ON THE INTERMEDIATE EXAMINATIONS

The scale is from 1 (lowest) to 6 (highest).

Subject	Einstein	Marić
	1898	1899
Differential & Integral Calculus	5.5	5
Analytic Geometry	6	5
Descriptive Geometry & Geometry of Position	5.5	4.75
Mechanics	6	5
Physics	5.5	5.5
Average Grade	5.7	5.05

Sources: For Marić, Zimmermann 1988a, 63; for Einstein, *CPAE* 1, doc. 42.

GRADES ON THE FINAL DIPLOMA EXAMINATIONS

Grades range from 1 to 6 (highest), but they are weighted. Grades doubled for theoretical physics, practical physics, and theory of functions. Grade for thesis is quadrupled, and astronomy is left as is. The average is total points divided by 11.

Subject	Einstein	Marić 1	Marić 2
	1900	1900	1901
Theoretical Physics	10	9	8
Practical Physics	10	10	8
Theory of Functions	11	5	7
Astronomy	5	4	5
Diploma Thesis	18	16	16
Total points	54	44	44
Average grade	4.91	4.00	4.00

Sources: *CPAE* 1, doc. 67; and Minkowski 1901 (see document list in bibliography).

APPENDIX E
GRADES ON LEAVING CERTIFICATES

The scale is from 1 (lowest) to 6 (highest). Fractional grades are rendered as decimals.

Subject	Einstein	Marić
	1900	1900
Differential & Integral Calculus with Exercises	4.75	4.5
Differential Equations with Exercises	5	
Descriptive Geometry with Exercises	4.25	4.25
Projective Geometry	4.25	3.5
Geometry of Position		5
Mechanics with Exercises	5.25	4.5
Analytic Geometry	5	4.5
Physics Practicum for Beginners	1	5
Weber, Physics	5.25	5
Electrotechnical Laboratory	6	
Weber, Scientific Projects in Physics Laboratories	5.33	5.33
Determination of Geographical Location	4.5	5

Sources: For Einstein, *CPAE* 1, doc. 28, p. 49; for Marić, Student Record, and facsimile T-G 1983, 57; T-G 1988, 61.

NOTES

INTRODUCTION

1. The speakers at the AAAS session were Robert S. Cohen, Caroline L. Herzenberg, Ruth H. Howes, Lewis R. Pyenson, John J. Stachel, Senta Troemel-Ploetz, and Evan Harris Walker (Herschman 1989, 1326).

CHAPTER 1: TWO TRAJECTORIES

1. In contrast, Abraham Fraenkel, a famed mathematician and the first dean of mathematics at The Hebrew University of Jerusalem, attended the Luitpold Gymnasium a short time after Einstein and later referred to the "nine happy years" that he spent there (Fraenkel 1955, 16; quoted in Fölsing 1997, 19).

CHAPTER 2: THE ZURICH POLYTECHNIC

1. According to another unverified hearsay report (Kleinknecht 2017, 178), Lieserl was adopted by a farm family in southern Germany and died there in 1980.

CHAPTER 4: WOMEN IN SCIENCE

1. Findlen 1993; Zinsser 2006; Ogilvie and Harvey 2000, "Sophie Germain (1776-1831)," vol. 1, 495–497; Gray 1987; Ogilvie and

Harvey 2000, "Mary (Fairfax) Greig Somerville (1780–1872)," vol. 2, 1213–1215; Koblitz 1983; Osen 1974.

2. Abir-Am and Outram 1987.

3. Ogilvie and Harvey 2000, "Caroline Lucretia Herschel (1750–1848)," vol. 1, 587–589.

4. Kohlstedt 1987; Ogilvie and Harvey 2000, "Maria Mitchell (1818–1889)," vol. 2, 901–905.

5. Jacques-Louis David, 1788, *Portrait of Antoine-Laurent Lavoisier and His Wife* (oil on canvas), Metropolitan Museum of Art, New York. See also Ogilvie and Harvey 2000, "Marie Anne Pierrette Paulze Lavoisier (1758–1836)," vol. 2, 752–753; Hoffmann 2002.

6. Rossiter 1982; Keller 1985, chapter 4; Abir-Am and Outram 1987, 2–6; Kohlstedt and Longino 1997.

7. Rossiter 1982, chaps. 8 and 9.

8. Rossiter 1982, chap. 1.

9. Citing "hired by observatories": Ogilvie and Harvey 2000, "Annie Scott Dill Russell Maunder (1868–1947)," vol. 2, 855–856; Sobel 2016; Vuong 2015. Citing "at laboratories," at ATT's Bell Labs, for example: https://blogs.scientificamerican.com/voices/betty-shannon-unsung-mathematical-genius/ (accessed August 16, 2018). Citing "government facilities," Langley Memorial Aeronautical Laboratory: https://www.nasa.gov/feature/when-the-computer-wore-a-skirt-langley-s-computers-1935-1970 (accessed August 16, 2018); for Jet Propulsion Lab: https://www.theatlantic.com/science/archive/2016/06/the-women-behind-the-jet-propulsion-laboratory/482847/ (accessed August 16, 2018). Citing "segregated by gender and at times by race": Shetterly 2016; Melfi 2016.

10. Howes and Herzenberg 1999.

11. Citing "radioactivity: Rayner-Canham and Rayner-Canham 1997; Rentetzi 2004. Citing "X-ray crystallography": Ferry 2014. Citing "astronomy": Kidwell 1984; Rossiter 1982, 53–57.

12. Pycior, Slack, and Abir-Am 1996.

13. Pycior 1987 and 1996; Quinn 1995.

14. McGrayne 1993, 93–116; Cohn 1996; Ogilvie and Harvey 2000, "Gerty Theresa Radnitz Cori (1896–1957)," 292–294.

15. Dash 1973, 227–346; Johnson 1986; McGrayne 1993, 175–200; Mozkowski 2006.

16. Citing "suicide as protest": Leitner 1994. Citing "despondent" and "loss of identity": Friedrich and Hoffmann 2017.

17. Dick 1981; Byers 2006; McGrayne 1993, 64–89.

18. Galison 1997, chap. 3; Halpern and Shapiro 2006; Strohmaier and Rosner 2006; Sime 2013.

19. Sime 1996 and 2005; Crawford, Sime, and Walker 1997.

20. Quinn 1995, chaps.13 and 14.

21. Summers 2005; see also http://www.thecrimson.com/article/2005/2/17/summers-releases-transcript-of-remarks-on (accessed August 16, 2018).

CHAPTER 5: THE STORY BEGINS

1. There is no consensus about the publication, unless the interview was published in two publications on the same date. The two contenders are *Politika* (T-G 1983, 75; Krstić 2004, 232) and *Vreme* (Time) (Zackheim 2000, 318; Popović, as editor of M-KS, 7; Milentijević 2015, 443). In no case did the authors/editor include the page numbers.

2. The translation of the last sentence of this paragraph in M-KS (omitted here) is incoherent. Comparison of the original German in Popović (1998, 288) with the translation in Dord Krstić's 2004 book shows that the latter is much more accurate. It reads: "She has written to me that way, and I let it be accepted that way, for otherwise this whole thing would be nonsense" (Krstić 2004, 197–198).

3. For more on the offer to Einstein and his acceptance of the Prague professorship in theoretical physics, see Fölsing 1997, 269–275; Isaacson 2007, 162–168.

CHAPTER 6: THE STORY EMERGES

1. Troemel-Ploetz did not trust the subsequent German editions of Trbuhović-Gjurić's book, since the editor had updated those editions with his own remarks following the publication of *CPAE* volume 1 (T-P 1990, 416–417).

2. This information was provided to Isaacson by Barbara Wolff of the Albert Einstein Archives at The Hebrew University of Jerusalem following the release of a new batch of letters on July 8, 2006, that had been embargoed by Einstein's stepdaughter, Margot, until the twentieth anniversary of her death. See Isaacson 2007, 606–607, n. 15, for the full story.

CHAPTER 7: COLLABORATION AS STUDENTS

1. In fall 1901, after she had learned she was pregnant, Marić wrote in desperation to Kaufler Savić: "Oh, Helene, pray to St. Peter for me that I might have him completely, that I do not have to be parted from him all the time—I love him so frightfully" (M-KS, 77).

CHAPTER 8: COLLABORATION DURING THEIR MARRIAGE

1. Much of what follows is based on a comprehensive analysis of the Joffe story by John Stachel (2005, liv–lxiii).

2. The convoluted arguments and ill-informed assertions made by Walker in his 1990 AAAS talk were decisively rebutted by Stachel in his response at that same meeting. See Stachel 2002, 31–38.

3. *CPAE* 5, docs. 48, 54, 56, 69, 86, 104, 108, 122, 124, 134, 150, 177, 190, 198, 202, 332).

4. *Patent and Trademark Library Association Newsletter* 15 (2004–2005): 40.

5. I contacted Michele Zackheim (August 10, 2010) to ascertain what evidence she had for stating that Einstein did not accompany Marić

when she visited Novi Sad in late summer 1907. She replied (August 10, 2010) that at that time she was not in a position to consult her very extensive notes from her three extended visits to Serbia in the 1990s.

6. In her *Reminiscences*, Elizabeth Roboz Einstein, the widow of Hans Albert, includes a photo of Hans Albert as a toddler together with his parents against a neutral background with the caption "The Einsteins in Novi Sad 1907 or 1908" (E. R. Einstein 1991, 8). However, there is no evidence that the whole family visited Novi Sad in 1907 or 1908—Zackheim, in her caption to the same photo, states that it was "taken in Bern probably around 1906."

7. A number of books by Forsee were published by Viking Children's Books and Macrae-Smith, a publisher of books for children and young readers.

8. Forsee 1963, 6–15, 18, 19, 22–23, 25, 30, 33–34, 36–37, 40, 41–43, 44–49, 55–58, 60–63, etc.

9. It is now a privately owned museum. But there is no mention of this quotation in the documentary account of Einstein in Bern by the museum's founder and curator Max Flückiger (1974).

CHAPTER 9: THE STORY SPREADS

1. See, for example, the extensive bibliography "Women in the Scientific Professions" by the Office of the Gender and Women's Studies Librarian, Libraries of the University of Wisconsin-Madison, at: https://www.library.wisc.edu/gwslibrarian/publications/bibliographies/science/professions/ (accessed August 14, 2018).

2. Hahn's Nobel Prize should also have been shared with Fritz Strassmann, Hahn's collaborator on the crucial experiment.

3. For example, Conway 1995.

4. Since he was some two years below the stipulated minimum age for taking the entrance examinations, Einstein obtained a special dispensation from the director of the Polytechnic, Professor Albin Herzog, who granted it after receiving documentary evidence of Einstein's exceptional prowess in mathematics (Fölsing 1997, 36–37, 36, n. 15).

5. This statement has since been removed from the Japan Prize website.

6. "Letter from Web Site Editor Andrea Gabor, September 24, 2007." No longer available on the PBS website.

7. Krstić (2004, 156) writes that as late as 1911, Marić "still lived for science."

CHAPTER 10: THE STORY CONTINUES

1. Milentijević provides full details of the Divorce Agreement and Divorce Decree provisions later in her book in their appropriate chronological places (Milentijević 2015, 235, 252).

2. To simplify these stipulations, the matter of a separate 40,000 marks, originally provided for Marić as alimony prior to Einstein's being awarded the Nobel Prize for Physics in 1922, has been omitted. See the fuller discussion in chapter 6. In references to the "Nobel Prize" in this document, it is evident from the context that the reference is to the financial award.

3. The letters Marić wrote in reply to Einstein's have not survived.

4. The letters in question are in the Albert Einstein Archives, The Hebrew University of Jerusalem, which is where Milentijević examined them. They were recently published in *CPAE* 15 (docs. 88, 95, 99) in the original German with an accompanying English translation volume.

5. Einstein refers to it only as "my will" (*CPAE* 15, Engl., doc. 79).

6. Zackheim suggests that Marić was considering writing her memoirs in order to make money (Zackheim 2000, 187).

7. Prior to its publication, Einstein's friends were horrified at the prospect of a book by an author previously known mainly through his books on jokes and the occult, and which was advertised as being based on conversations between Einstein and Moszkowski. They feared it would provide ammunition for the campaign against Einstein being waged by anti-Semites in Germany at that time (Fölsing 1997, 469).

8. The editors of *CPAE* 11 (p. 186) state that Laub stayed in Bern for three weeks in April 1908, but there is good evidence that Laub actually stayed for at least five weeks, extending into May (*CPAE* 5, docs. 91, 96, 101; Fölsing 1997, 240 and 240, n. 25). See also "Einstein and Laub on the electrodynamics of moving bodies" (*CPAE* 2, ed. note, 503–507).

BIBLIOGRAPHY

All websites accessible as of October 15, 2018

ARCHIVAL DOCUMENTS

Nobel Prize. Nomination Archive. https://www.nobelprize.org/nomination/archive/.

Minkowski, Hermann. 1901. Report. ETH Section VI A, "Protokoll der Abteilungs-Sitzungen (Minutes of Meetings)." Meeting on July 26, 1901, ETH Institute Archive, Zurich. Hs. 1079–02–1901–07–26. Results of diploma exams on July 16, 1901.

Marić, Mileva. Student Record. Course and Grade Report (*Matrikel*) with Leaving Certificate (*Abgangszeugnis*). ETH Library, Zurich. Archives und Papers (*Nachlässe*). RA_Matrikel-85-Maric.pdf. Online at http://www.library.ethz.ch/en%20/Resources/Digital-library/Einstein-Online/Einstein-s-Studies-at-the-Polytechnic-Institute-in-Zurich-1896-1900.

PUBLICATIONS

Abir-Am, Pnina, and Dorinda Outram, eds. 1987. *Uneasy Careers, Intimate Lives. Women in Science, 1789–1978*. New Brunswick, NJ: Rutgers University Press.

Alter, Svetlana. 2013. *Secret Traces of the Soul of Mileva Marić-Einstein*. Pittsburgh: Dorrance Publishing.

Arianrhod, Robyn. 2014. *Young Einstein and the Story of E=mc²*. Hampress, Amazon Kindle Singles.

Barnett, Carol. 1998. "A Comparative Analysis of Perspectives of Mileva Marić Einstein." Master of Arts thesis. Florida State University.

Beyerchen, Alan. 1977. *Scientists Under Hitler: Politics and the Physics Community in the Third Reich*. New Haven: Yale University Press.

Benedict, Marie. 2016. *The Other Einstein: A Novel*. Naperville, IL: Sourcebooks Landmark.

Borchardt, Edith. 2008. "Introduction." In Chiu 2008, 1–27.

Bernstein, Jeremy. 1998. "The Road to E = mc^2." *Times Higher Educational Supplement* 25 (September).

Bjerknes, Christopher. 2002. *Albert Einstein: The Incorrigible Plagiarist*. Downers Grove, IL: XTX.

Byers, Nina. 2006. "Emmy Noether (1882–1935)." In Byers and Williams 2006, 83–96.

Byers, Nina and Gary Williams. 2006. *Out of the Shadows: Contributions of Twentieth-Century Women to Physics*. New York: Cambridge University Press.

Calaprice, Alice, Daniel Kennefick, and Robert Schulmann. 2015. *An Einstein Encyclopedia*. Princeton, NJ: Princeton University Press.

Cassidy, David C. 2004. *Einstein and Our World*. 2nd ed. Amherst, NY: Humanity Books.

Chiu, Charles S. 2008. *Women in the Shadows*. Edith Borchardt, trans. New York: Peter Lang.

Clark, Linda L. 2008. *Women and Achievement in Nineteenth-Century Europe*. New York: Cambridge University Press.

Clark, Ronald. 1971. *Einstein: The Life and Times*. New York: World Publishing Company.

Cohn, Mildred. 1996. "Carl and Gerty Cori. A Personal Recollection." In Pycior, Slack, and Abir-An 1996, 72–84.

Conway, Jill Ker. 1995. Review of Gabor 1995. *New York Times*, October 8.

Crawford, Elisabeth, Ruth Lewin Sime, and Mark Walker. 1997. "A Postwar Tale of Nobel Injustice." *Physics Today* 50, no. 9 (September): 26–32.

Čvorić, Mirjana. 2006. *One Stone—Einstein.* Production TV Šabac, Serbia (English subtitles). http://www.youtube.com/watch?v=htfBTfVD75o.

Danin, Daniil Semenovich. 1962. *Neizbezhnost strannogo mira*, Moscow: Molodaia Gvardia, Gosudatsvenaaja Bilblioteka SSSR.

Dash, Joan. 1973. *A Life of One's Own. Three Gifted Women and the Men They Married.* New York: Harper & Row.

de Vrieze, Jop. 2017. "'Science Wars' Veteran Has a New Mission." *Science* 358 (October 13): 159. Interview with Bruno Latour.

Dick, Auguste. 1981. *Emmy Noether 1882–1935.* H. I. Blocher, transl. Boston: Birkhäuser.

Eckart, Wolfgang U. 2001. "Zunächst jedoch nur versuchs- und probeweise"–vor 100 Jahren: Die ersten Medizinstudentinnen beziehen die Universität Heidelberg. http://www.uni-heidelberg.de/institute/fak5/sonstiges/timeline/Frau_hd.pdf.

Edelson, Micah, et al. 2011. "Following the Crowd: Brain Substrates of Long-Term Memory Conformity." *Science* 333 (July): 108–111.

Einstein, Albert. 1949. "Autobiographical Notes." English with parallel German text. In *Albert Einstein, Philosopher-Scientist.* P. A. Schilpp, ed. and trans. Part 1, 1–95. La Salle, IL: Open Court. Reprinted Einstein 1979.

Einstein, Albert. 1955. "Erinnerungen-Souvenirs." *Schweizerische Hochschulzeitung* 25 (*Sonderheft*): 145–153. Reprinted as Einstein 1956a.

Einstein, Albert. 1956a. "Autobiographische Skizze." Reprint of Einstein 1955. In Seelig 1956b, 9–17.

Einstein, Albert. 1956b. *Lettres à Maurice Solovine.* Maurice Solovine, French transl. Paris: Gauthier-Villars.

Einstein, Albert. 1979. *Autobiographical Notes*. La Salle, IL: Open Court. Reprint of Einstein 1949.

Einstein, Albert. 1987ff. *The Collected Papers of Albert Einstein*. Vols. 1–15 (currently), with accompanying English translations for each volume. Various editors. Princeton, NJ: Princeton University Press. Fully accessible online, along with English translations, at: http://einsteinpapers.press.princeton.edu/. Abbreviated *CPAE*.

Einstein, Albert, and Mileva Marić. 1992. *The Love Letters*. Jürgen Renn and Robert Schulmann, eds. Shawn Smith, trans. Princeton, NJ: Princeton University Press. Abbreviated E-M.

Einstein, Albert, and Arnold Sommerfeld. 1968. *Briefwechsel: Sechzig Briefe aus dem goldenen Zeitalter der modernen Physik*. Armin Hermann, ed. Basel: Schwabe & Co.–Verlag.

Einstein, Elizabeth Roboz. 1991. *Hans Albert Einstein: Reminiscences of His Life and Our Life Together*. Iowa City: Iowa Institute of Hydraulic Research, University of Iowa.

Esterson, Allen. 2006a. "*Einstein's Wife* Documentary: List of Errors." http://www.esterson.org/Einsteins_Wife_Errors_List.htm.

Esterson, Allen. 2006b. "*Einstein's Wife*: Mileva Marić 1." http://www.esterson.org/einsteinwife1.htm. (Statement to PBS).

Esterson, Allen. 2006c. "Critique of Evan Harris Walker's Letter in *Physics Today* February 1991". http://www.esterson.org/Walker_Physics_Today.htm.

Fernyhough, Charles. 2012. *Pieces of Light: The New Science of Memory*. London: Profile Books.

Ferry, Georgina. 2014. "Women in Crystallography." *Nature* 505 (January): 609–611.

Findlen, Paula. 1993. "Science as a Career in Enlightenment Italy. The Strategies of Laura Bassi." *Isis* 84: 441–469.

Flückiger, Max. 1974. *Albert Einstein in Bern: Das Ringen um ein neues Weltbild: Eine dokumentarische Darstellung über den Aufstieg eines Genies*. Bern: Verlag Paul Haupt.

Fölsing, Albrecht. 1990. "Keine 'Mutter der Relativitätstheorie.'" *Die Zeit* 47 (November 16). Translation by Josephine Riches, at http://www.esterson.org/Foelsing_Die_Zeit_1990.htm.

Fölsing, Albrecht. 1997. *Albert Einstein.* Ewald Osers, trans. New York: Penguin Books.

Forsee, Aylesa. 1963. *Albert Einstein: Theoretical Physicist.* New York: Macmillan.

Fraenkel, Abraham A. 1955. *Lebenskreise—Erinnerungen eines jüdischen Mathematikers.* Stuttgart: Deutsche Verlags-Anstalt.

Frank, Philipp. 1947. *Einstein: His Life and Times.* Shuichi Kusaka, ed. George Rosen, trans. New York: A. A. Knopf.

Friedrich, Bretislav, and Dieter Hoffmann. 2017. "Clara Immerwahr. A Life in the Shadow of Fritz Haber." In *One Hundred Years of Chemical Warfare. Research, Deployment, Consequences.* Bretislav Friedrich, Dieter Hoffmann, Jürgen Renn, Florian Schmaltz, and Martin Wolf, eds., 45–67 Springer Open eBook.

Frize, Peter. 2009. Chapter 15. "The Bold and the Brave: Sophie Germaine, Mileva Marić -Einstein and Rosalind Franklin." In Monique Frize. *The Bold and the Brave: A History of Women in Science and Engineering,* 261–298. Ottawa: University of Ottawa Press.

Gabor, Andrea. 1995. "Mileva Marić Einstein." In *Einstein's Wife: Work and Marriage in the Lives of Five Great Twentieth-Century Women,* 1–32. New York: Viking.

Galison, Peter. 1987. *How Experiments End.* Chicago: University of Chicago Press.

Galison, Peter. 1997. *Image and Logic: A Material Culture of Microphysics.* Chicago: University of Chicago Press.

Galison, Peter. 2003. *Einstein's Clocks and Poincaré's Maps: Empires of Time.* New York: W. W. Norton.

Gardner, Martin. 1981. "Parapsychology and Quantum Mechanics." In *Science and the Paranormal: Probing the Existence of the Supernatural.* George O. Abell and Barry Singer, eds., 56–69. New York: Scribner.

Getler, Michael. 2006. "Einstein's Wife: The Relative Motion of 'Facts.'" PBS Ombudsman Column, December 15, 2006. http://www.pbs.org/ombudsman/2006/12/einsteins_wife_the_relative_motion_of_facts.html.

Goodman, Ellen. 1990. "Out of the Shadows of 'Great' Men." *Boston Globe*, March 15, 1990: 15 (op-ed column). Reprinted as "The Myth of Mileva: When It Came to Marriage, Einstein Was No Genius." *Chicago Tribune*, March 18, 1990.

Gornick, Vivian. 2009. *Women in Science: Then and Now*. New York: Feminist Press at the City University of New York.

Gray, Mary W. 1987. "Sophie Germain (1776–1831)." In *Women of Mathematics. A Biobibliographic Sourcebook*. Louise S. Grinstein and Paul Campbell, eds., 47–55. Westport, CT: Greenwood Press.

Gutfreund, Hanoch, and Jürgen Renn. 2015. *The Road to Relativity: The History and Meaning of Einstein's "The Foundation of General Relativity."* Princeton, NJ: Princeton University Press.

Habicht, Conrad, and Paul Habicht. 1910. "Elektrostatischer Potentialmultiplikator nach A. Einstein." *Physikalische Zeitschrift* 11: 532–535.

Halpern, Leopold, and Maurice M. Shapiro. 2006. "Marietta Blau (1894–1970)." In Byers and Williams 2006, 109–126.

Hentschel, Ann, and Gerd Grasshoff. 2005. *Albert Einstein: "Those Happy Bernese Years."* Bern: University of Bern.

Hentschel, Klaus 1990. *Interpretationen und Fehlinterpretationen der speziellen und allgemeinen Relativitätstheorien durch Zeitgenossen Albert Einsteins*. Basel: Birkhäuser.

Herschman, Arthur. 1989. "1990 AAAS Annual Meeting, New Orleans, 15-20 February." *Science* 246: 1313–1327. Meeting program.

Hertz, Heinrich. 1892. *Untersuchungen über die Ausbreitung der elektrischen Kraft*. Leipzig: Barth.

Highfield, Roger, and Paul Carter. 1993. *The Private Lives of Albert Einstein*. London: Faber and Faber.

Hilton, Geraldine. 2003. *Einstein's Wife: The Life of Mileva Marić Einstein*. Paul Humfress, exec. producer. Film produced by Melsa Films in conjunction with Oregon Public Broadcasting and Australian Broadcasting Corporation.

Hoffmann, Dieter. 2013. *Einstein's Berlin: In the Footsteps of a Genius*. Baltimore: Johns Hopkins University Press.

Hoffmann, Roald. 2002. "Mme. Lavoisier." *American Scientist* 90 (Jan.-Feb.): 22–24.

Holton, Gerald. 2000. "Of Physics, Love, and Other Passions: The Letters of Albert and Mileva." In *Einstein, History, and Other Passions: The Rebellion against Science at the End of the Twentieth Century*. 2nd ed., 170–193. Cambridge, MA: Harvard University Press.

Howard, Ron et al., producers. 2017. *Genius: Einstein*. Television series, 10 episodes. Developed for TV by Noah Pink and Ken Biller. National Geographic Channel.

Howes, Ruth H., and Caroline L. Herzenberg. 1999. *Their Day in the Sun. Women of the Manhattan Project*. Philadelphia: Temple University Press.

Hunter, Ian Melville Logan. 1957. *Memory: Facts and Fallacies*. Harmondsworth, Middlesex: Penguin Books.

Illić, Mirjana, and Andreas Kleinert. 2003. "'Allerliebstes Helenchen': Mileva Einsteins Briefe an Helene Savić." *NTM* 11: 29–33.

Irons, F. E. 2004. "Reappraising Einstein's 1909 Application of Fluctuation Theory to Planckian Radiation." *American Journal of Physics* 72: 1059–1067. http://users.df.uba.ar/giribet/f4/einstein_planck.pdf.

Isaacson, Walter. 2007. *Einstein: His Life and Universe*. New York: Simon & Schuster.

Janssen, Michel, and Jürgen Renn. 2015. "Arch and Scaffold: How Einstein Found His Field Equations." *Physics Today* 68, no. 11: 30–36.

Jarausch, Konrad H. 1982. *Students, Society, and Politics in Imperial Germany: The Rise of Academic Illiberalism*. Princeton, NJ: Princeton University Press.

Joffe, Abraham F. 1955. "Pamyati Alberta Eynshtyna" [Memorial for Albert Einstein]. *Uspekhi fizicheskikh nauk* 57, no. 2: 188-192.

Joffe, Abraham F. 1962. *Vstrechi s fizikami, moi vospominaniia o zarubezhnykh fizikah* [Meetings with Physicists, My Reminiscences of Physics Abroad]. Moscow: Gusudarstvennoye Izdatelstvo Fiziko-Matematitsheskoi Literatury. For German translation see Joffe 1967.

Joffe, Abraham F. 1967. *Begegnungen mit Physikern*. Konrad Werner, trans. Leipzig: Teubner Verlag. Translation of Joffe 1962.

Johnson, Karen E. 1986. "Maria Goeppert-Mayer. Atoms, Molecules, and Nuclear Shells." *Physics Today* 39, no. 9 (September): 44–49.

Johnson, M. Alex. 2005. "The Culture of Einstein." NBC News, April 18, 2005. http://www.nbcnews.com/id/7406337/#.Upyg ZOK7SOt.

Jungnickel, Christa, and Russell McCormmach. 1986. *Intellectual Mastery of Nature*. Vol. 2. Chicago: University of Chicago Press.

Južnić, Stanislav. 2004. "Appendix C." In Krstić 2004, 236–246.

Kaiser, David. 2011. *How the Hippies Saved Physics: Science, Counterculture, and the Quantum Revival*. New York: W. W. Norton & Company.

Keller, Evelyn Fox. 1985. Chapter 4. "Gender in Science." In *Reflections on Gender and Science*. New Haven: Yale University Press.

Kidwell, Peggy Aldrich. 1984. "Women Astronomers in Britain 1780–1930." *Isis* 75: 534–546.

Kien, Jenny and D. C. Cassidy. 1984. "The History of Women in Science: A Seminar at the University of Regensburg, FRG." *Women's Studies International Forum* 7: 313–317.

Klein, Martin J. 1963. "Einstein's First Paper on Quanta." *The Natural Philosopher* 2: 59–86.

Kleinert, Andreas. 2005. Review of Popović 2003. *NTM* 13: 125–126.

Kleinknecht, Konrad. 2017. *Einstein und Heisenberg: Begründer der modernen Physik*. Stuttgart: Kohlhammer.

Koblitz, Anne Hibner. 1983. *A Convergence of Lives. Sofia Kovalevs-kaia, Scientist, Writer, Revolutionary.* Boston: Birkhäuser.

Kohlstedt, Sally Gregory. 1987. "Maria Mitchell and the Advancement of Women in Science." In Abir-Am and Outram 1987, 129–146.

Kohlstedt, Sally Gregory, and Helen E. Longino, eds. 1997. *Women, Gender, and Science. New Directions (Osiris* 12). Chicago: University of Chicago Press.

Kollros, Louis. 1956. "Erinnerungen eines Kommilitonen." In Seelig 1956b, 17–31.

Krstić, Dord. 1991. "Appendix A: Mileva Einstein-Marić." In E. R. Einstein 1991, 85–99.

Krstić, Dord. 2004. *Mileva & Albert Einstein: Their Love and Scientific Collaboration.* Kranjska, Slovenia: Didakta.

Latour, Bruno. 1987. *Science in Action: How to Follow Scientists and Engineers Through Society.* Cambridge, MA: Harvard University Press.

Leitner, Gerit von. 1994. *Der Fall Clara Immerwahr. Leben für eine humane Wissenschaft.* Munich: Verlag C. H. Beck.

Lenard, Philipp. 1899. "Erzeugung von Kathodenstrahlen durch ultraviolettes Licht." *Wiener Berichte* 108: 1649–1666.

Lenard, Philipp. 1900. Re-publication of Lenard 1899. *Annalen der Physik* 2: 359–375.

Lenard, Philipp. 1902. "Über die lichtelektrische Wirkung." *Annalen der Physik* 8: 149–198.

Maas, Ad. 2007. "Einstein as Engineer: The Case of the Little Machine." *Physics in Perspective* 9: 305–328.

Marković, Živko. 1995. "Sećanje starih novosadana na Ajnštajnove" [Recollections of Old Novi Sad Residents about the Einsteins]. In *Dorpinos Mileve Ajinštajn-Marić nauci* [Contributions of Mileva Einstein-Marić to Science], Rastko Maglić, ed. Belgrade: MST Gajić.

Martínez, Alberto A. 2004. "Arguing About Einstein's Wife." *Physics World* 17 no. 4: 14.

Martínez, Alberto A. 2005. "Handling Evidence in History: The
Case of Einstein's Wife." *School Science Review* 86, no. 316 (March):
49–56.

Martínez, Alberto A. 2007. "Martínez Writings: Einstein's Wife."
http://martinezwritings.com/m/Maric.html.

Martínez, Alberto A. 2009. *Kinematics: The Lost Origins of Einstein's
Relativity.* Baltimore: Johns Hopkins University Press.

Martínez, Alberto A. 2011. *Science Secrets: The Truth About Darwin's
Finches, Einstein's Wife, and Other Myths.* Pittsburgh: University of
Pittsburgh Press.

Maurer, Margarete. 1996. "'Weil nicht sein kann was nicht sein darf'
… 'Die Eltern' oder 'der Vater' der Relativitätstheorie?" *PC News*
(Vienna) 48, vol. 11, no. 3 (June): 20–27.

McGrayne, Sharon Bertsch. 1993. *Nobel Prize Women in Science.
Their Lives, Struggles, and Momentous Discoveries.* New York: Birch
Lane Press.

Medicus, Heinrich, A. 1994. "The Friendship among Three Singular
Men: Einstein and His Swiss Friends Besso and Zangger." *Isis* 85:
456–478.

Melfi, Theodore (director). 2016. Film: *Hidden Figures.* Hollywood,
CA: 20th Century Fox.

Michelmore, Peter. 1962. *Einstein: Profile of the Man.* New York:
Dodd, Mead & Company; London: Frederick Muller, 1963.

Milentijević, Radmila. 2012. *Mileva Marić Ajnštajn: Život sa Alber-
tom Ajnštajnom.* Belgrade: Prosveta.

Milentijević, Radmila. 2013. *Mileva Marić Einstein: Vivre avec Albert
Einstein.* Paris: L'Age d'Homme. French edition of Milentijević 2012.

Milentijević, Radmila. 2015. *Mileva Marić Einstein: Life with Albert
Einstein.* New York: United World Press, distributed by BookBaby
Publishers and Amazon Kindle. English edition of Milentijević 2012.

Moszkowski, Alexander. 1921. *Einstein: Einblicke in seine Gedanken-
welt. Gemeinverständliche Betrachtungen über die Relativitätstheorie*

und ein neues Weltsystem. Hamburg and Berlin: Hoffmann und Campe.

Mozkowski, Stephen A. 2006. "Maria Goeppert-Mayer (1906–1972)." In Byers and Williams 2006, 202–212.

Neffe, Jürgen. 2007. *Einstein: A Biography.* Shelley Frisch, trans. New York: Farrar, Straus, Giroux.

New York Times. 1990. "Did Einstein's Wife Contribute to His Theories?" *New York Times,* March 27, 1990.

Ogilvie, Marilyn, and Joy Harvey, eds. 2000. *The Biographical Dictionary of Women in Science. Pioneering Lives from Ancient Times to the Mid-Twentieth Century.* New York: Routledge.

Osen, Lynn M. 1974. *Women in Mathematics.* Cambridge, MA: MIT Press.

Overbye, Dennis. 2001. *Einstein in Love: A Scientific Romance.* New York: Penguin.

Pais, Abraham. 1982. *"Subtle is the Lord ..." The Science and the Life of Albert Einstein.* New York: Oxford University Press.

Pais, Abraham. 1994. *Einstein Lived Here.* New York: Oxford University Press.

Palmer, R. P., Joel Colton, and Lloyd Kramer. 2007. *A History of the Modern World.* 10th ed. New York: McGraw-Hill College.

Planck, Max. 1910. "Zur Theorie der Wärmestrahlung." *Annalen der Physik* 31: 758–768.

Plous, Scott. 1993. *The Psychology of Judgment and Decision Making.* New York: McGraw-Hill.

Popović, Milan, ed. 1998. *Jedno prijateljstvo: pisma Mileve i Alberta Ajnštajna Heleni Savić* [A Friendship: Letters from Mileva and Albert Einstein to Helene Savić]. Podgorica, Montenegro: CID. Marić's original German letters with parallel Serbian translation.

Popović, Milan, ed. 2003. *In Albert's Shadow: The Life and Letters of Mileva Marić, Einstein's First Wife.* Translation of letters in Popović 1998. Translator not indicated. Baltimore: Johns Hopkins University Press.

Pycior, Helena M. 1987. "Marie Curie's 'Anti-natural Path.' Time Only for Science and Family." In Abir-Am and Outram 1987, 191–214.

Pycior, Helena M. 1996. "Pierre Curie and 'His Eminent Collaborator Mme. Curie.'" In Pycior, Slack, and Abir-Am 1996, 39–56.

Pycior, Helena M., Nancy G. Slack, and Pnina G. Abir-Am, eds. 1996. *Creative Couples in the Sciences (Lives of Women in Science)*. New Brunswick, NJ: Rutgers University Press.

Pyenson, Lewis. 1985. *The Young Einstein: The Advent of Relativity*. Bristol, UK: Adam Hilger.

Quinn, Susan. 1995. *Marie Curie: A Life*. New York and London: Simon & Schuster.

Rayner-Canham, Marelene F., and Geoffrey W. Rayner-Canham, eds. 1997. *A Devotion to Their Science. Pioneer Women of Radioactivity*. Philadelphia: Chemical Heritage Foundation; Montreal & Kingston: McGill-Queen's University Press.

Renn, Jürgen et al. 2007. *The Genesis of General Relativity (Boston Studies in the Philosophy of Science)*. 4 vols. New York: Springer Publishing.

Rentetzi, Maria. 2004. "Gender, Politics, and Radioactivity Research in Interwar Vienna. The Case of the Institute for Radium Research." *Isis* 75: 359–393.

Rose, Hilary. 1994. *Love, Power and Knowledge: Towards a Feminist Transformation of the Sciences*. Cambridge: Polity Press; Bloomington: Indiana University Press.

Rossiter, Margaret W. 1982. *Women Scientists in America: Struggles and Strategies to 1940*. Baltimore: Johns Hopkins University Press.

Rossiter, Margaret W. 1995. *Women Scientists in America: Before Affirmative Action 1940–1972*. Baltimore: Johns Hopkins University Press.

Rossiter, Margaret W. 2012. *Women Scientists in America: Forging a New World since 1972*. Baltimore: Johns Hopkins University Press.

Schulmann, Robert, and Gerald Holton. 1995. "Einstein's Wife." Letter to the editor. *New York Times*, October 8.

Seelig, Carl. 1954. *Albert Einstein, eine Dokumentarische Biographie.* 2nd ed. Zurich: Europa Verlag..

Seelig, Carl. 1956a. *Albert Einstein: A Documentary Biography.* Translation of Seelig 1954. London: Staples Press.

Seelig, Carl, ed. 1956b. *Helle Zeit—Dunkle Zeit: In Memoriam Albert Einstein.* Zurich: Europa Verlag. Reissued 1986 with an introduction by Karl von Meyenn.

Shetterly, Margot Lee. 2016. *Hidden Figures. The Untold Story of Four African-American Women Who Helped Launch Our Nation into Space.* New York: William Morrow & Co.

Sime, Ruth Lewin. 1996. *Lise Meitner: A Life in Physics.* Berkeley: University of California Press.

Sime, Ruth Lewin. 1997. "Unburying Treasure." *Barnard*, Spring: 56.

Sime, Ruth Lewin. 2005. "From Exceptional Prominence to Prominent Exception: Lise Meitner at the Kaiser Wilhelm Institute for Chemistry." In *Forschungsprogramm "Geschichte der Max-Planck-Gesellschaft im Nationalsozialismus,"* *Ergebnisse 24*. Susanne Heim, ed. Berlin: Max-Planck-Gesellschaft.

Sime, Ruth Lewin. 2013. "Marietta Blau: Pioneer of Photographic Nuclear Emulsions and Particle Physics." *Physics in Perspective* 15: 3–32.

Smith, Dinitia. 1996. "Dark Side of Einstein Emerges in His Letters." *New York Times*, November 6.

Sobel, Dava. 2016. *The Glass Universe. How the Ladies of the Harvard Observatory took the Measure of the Stars.* New York: Penguin Books.

Soden, Kristine von, and Gaby Zipfel, eds. 1979. *70 Jahre Frauenstudium: Frauen in der Wissenschaft.* Cologne: Pahl-Rugenstein.

Solovine, Maurice. 1956. "Introduction." In Einstein 1956b, v-xiii.

Solovine, Maurice. 1959. "Freundschaft mit Albert Einstein." *Physikalische Blätter* 15, no. 3: 97–103.

Sretenović, Miša. 1929. "Ljubavni roman slavnog naučnika Alberta Ajnštajna i Srpkinje, Mileve Marić." *Politika* (Belgrade), May 23, 1929, or *Vreme*, May 23, 1929. No page numbers given in any source.

Stachel, John. 2002. *Einstein from 'B' to 'Z'* (*Einstein Studies*, vol. 9). Boston: Birkhäuser.

Stachel, John. 2005. "Introduction to the Centenary Edition." In *Einstein's Miraculous Year: Five Papers That Changed the Face of Physics*. John Stachel, ed., xv-lxxii. Princeton, NJ: Princeton University Press. Originally published 1998.

Strohmaier, Brigitte, and Robert Rosner, eds. 2006. Paul F. Dvorak, ed. English edition. *Marietta Blau—Stars of Disintegration. Biography of a Pioneer of Particle Physics*. Riverside, CA: Ariadne Press.

Summers, Lawrence H. 2005. "Remarks at National Bureau of Economic Research Conference, January 14, 2005." https://www.harvard.edu/president/speeches/summers_2005/nber.php.

Talmey, Max. 1932. *The Relativity Theory Simplified and the Formative Period of Its Inventor*. New York: Falcon Press.

Trbuhović-Gjurić [also Đjurić], Desanka. 1969. *U senci Alberta Ajnštajna*. Kruševac, then-Yugoslavia: Bagdala.

Trbuhović-Gjurić, Desanka. 1982. *Mileva Marić-Einstein 1875–1948*. Translation of Trbuhović-Gjurić 1969. Translator not indicated. With a "Nachwort" by the editor, Werner G. Zimmermann, 187-189. Bern: Verlag Paul Haupt.

Trbuhović-Gjurić , Desanka. 1983. *Im Schatten Albert Einsteins: Das tragische Leben der Mileva Einstein- Marić*. 2nd edition of Trbuhović-Gjurić 1982. With Nachwort by Zimmermann, 187-189. Bern: Verlag Paul Haupt.

Trbuhović-Gjurić , Desanka. 1988. *Im Schatten Albert Einsteins: Das tragische Leben der Mileva Einstein- Marić*. New edition of Trbuhović-Gjurić 1983 with new material (see Zimmermann 1988a, 1988b). Bern: Verlag Paul Haupt.

Trbuhović-Gjurić , Desanka. 1991. *Mileva Einstein: Une Vie.* Translation of Trbuhović-Gjurić 1988 by Nicole Casanova. Paris: Antoinette Fouque (Editions des Femmes).

Trbuhović-Gjurić , Desanka. 1993. *Im Schatten Albert Einsteins: Das tragische Leben der Mileva Einstein- Marić.* Reissue of Trbuhović-Gjurić 1988. Bern: Verlag Paul Haupt.

Trbuhović-Gjurić [also Đjurić], Desanka. 1995. *U senci Alberta Ajnštajna.* Belgrade: Klub NT. Reprint of Trbuhović-Gjurić 1969.

Troemel-Ploetz, Senta. 1990. "Mileva Einstein-Marić: The Woman Who Did Einstein's Mathematics." *Women's Studies International Forum* 13, no. 5: 415–432. Abbreviated T-P 1990.

Vuong, Zen. 2015. "These Women Were 'Human Computers' Before They Were Allowed to Be Astronomers." *Pasadena Star-News.* Oct. 6. https://www.pasadenastarnews.com/2015/10/06/these -women-were-human-computers-before-they-were-allowed-to-be -astronomers/.

Walker, Evan H. 1979. "The Quantum Theory of Psi Phenomena." *Psychoenergetic Systems* 3: 259–299.

Walker, Evan H. 1989. "Did Einstein Espouse His Spouse's Ideas?" *Physics Today* 42, no. 2 (February): 9–11. Letter to the editor.

Walker, Evan H. 1990. "Ms. Einstein." Paper delivered at the AAAS session on "The Young Einstein," New Orleans, February 18, 1990. Unpublished.

Walker, Evan H. 1991. "Mileva Marić's Relativistic Role." *Physics Today* 44, no. 2 (February): 122–123. Letter to the editor.

Wazeck, Milena. 2014. *Einstein's Opponents: The Public Controversy about the Theory of Relativity in the 1920s.* New York: Cambridge University Press.

Whitrow, G. J., ed. 1967. *Einstein: The Man and His Achievement.* New York: Dover Publications.

Wolff, Barbara. 2016. "The Nobel Prize in Physics 1921–What Happened to the Prize Money?" http://www.einstein-website.de/z_ information/nobelprizemoney.html.

Zackheim, Michele. 2000. *Einstein's Daughter: The Search for Lieserl.* New York: Riverhead Books. The page numbering differs in the hardcover (1999) and paperback editions.

Zimmermann, Werner G. 1988a. "Mileva und Albert: Nachtrag des Herausgebers." In Trbuhović-Gjurić 1988, 59–78.

Zimmermann, Werner G. 1988b. "Nachwort." In Trbuhović-Gjurić 1988, 209–213. This differs from the "Nachwort" in Trbuhović-Gjurić 1982 and 1983, 187–189.

Zinsser, Judith P. 2006. *La Dame d'Esprit. A Biography of the Marquise du Châtelet.* New York: Viking.

INDEX